ELECTRICAL MACHINES ~~LAB~~

MATLAB® PROGRAMS

ELECTRICAL MACHINES LAB MANUAL

WITH

MATLAB® PROGRAMS

By

Dr. D.K. CHATURVEDI

Professor, Deptt. of Electrical Engineering
Faculty of Engineering, D.E.I. (Deemed University)
Dayalbagh, Agra (Uttar Pradesh)

UNIVERSITY SCIENCE PRESS

(An Imprint of Laxmi Publications Pvt. Ltd.)

BANGALORE ● CHENNAI ● COCHIN ● GUWAHATI ● HYDERABAD
JALANDHAR ● KOLKATA ● LUCKNOW ● MUMBAI ● PATNA
RANCHI ● NEW DELHI

Published by :

UNIVERSITY SCIENCE PRESS

(An Imprint of Laxmi Publications Pvt. Ltd.)
113, Golden House, Daryaganj,
New Delhi-110002
Phone : 011-43 53 25 00
Fax : 011-43 53 25 28

www.laxmipublications.com
info@laxmipublications.com

Price : $ 4.87 Only.

First Edition . 2010
Reprint : 2011

OFFICES

✆ Bangalore	080-26 75 69 30	✆ Chennai	044-24 34 47 26
✆ Cochin	0484-237 70 04, 405 13 03	✆ Guwahati	0361-251 36 69, 251 38 81
✆ Hyderabad	040-24 65 23 33	✆ Jalandhar	0181-222 12 72
✆ Kolkata	033-22 27 43 84	✆ Lucknow	0522-220 99 16
✆ Mumbai	022-24 91 54 15, 24 92 78 69	✆ Patna	0612-230 00 97
✆ Ranchi	0651-221 47 64		

UEM-9499-195-ELE MACH LAB MAN MATLAB-CHA   C—3830/011/07
Typeset at : Sukuvisa Enterprises, New Delhi.   *Printed at* : Ajit Printers, Delhi.

Dedicated to
The cherished memories of my Guru and Guide

Most Revered Dr. Makund Behari Lal Sahab
D.Sc. (Lucknow), D.Sc. (Edinburgh)
(1907 – 2002)
August Founder of Dayalbagh Educational Institute
Dayalbagh, Agra

CONTENTS

SECTION-F

PREFACE

Electrical machines and power system are the back bone of electrical engineering and play a vital role in industry. It is therefore essential that students of electrical engineering should have a proper background in these areas. It is observed that the students do not take much interest in the laboratory. Hence, the necessity is felt to explain the concepts of electrical machines with suitable computer programs besides the theory and practice. This will provide better understanding of the subject. There are certain limitations in lab experimentation that any arbitrary voltage or load to machines can not be applied, otherwise it will get damaged, but if a suitable Matlab® program consisting of machine model is available, then one may apply any voltage or load (*i.e.,* playing with the program) and learn in much better way.

Electrical Machines Lab Manual with Matlab® programs is a book for an alternate way of learning the subject. It explains the basic types of electrical machines, such as transformer, DC machine, induction machine, synchronous machine and low-power motors operated at single phase AC systems. It provides a basic knowledge of electrical machines along with the experimental details performed on these machines, useful for further studies and practice. To get different external and internal characteristics of these machines in the lab, systematic procedure is explained and the results are validated with Matlab® programs.

This book discusses the basics, structure and steady state operating behaviour, principle of operation of these machines, relevant theory, the procedure of performing experiments, observations and results analysis, Matlab® program codes and finally, to check the understanding multiple choice questions are also provided.

—**Author**

Chapter 1

INTRODUCTION

Lab work is very important to bridge the gap between the theoretical knowledge and practical work. The motto of the laboratory work is to learn here and earn outside.

Students always have fear while working experimentally in the laboratories. Especially in electrical engineering labs like machines lab where the voltage level is generally 230 V or 415 V at 50 Hz, which may cause a serious shock or sometimes cause death also. Hence, there is a need to follow certain precautions to work safely in the lab. While working in lab one must be properly isolated from ground (*i.e.,* wear proper shoes), so that there is no chance to flow the leakage current through the body.

1.1 SAFETY IN ELECTRICAL MACHINES LAB

Severity of the electrical shock depends on:

(*i*) *Path* of current through the body

(*ii*) *Amount of current* flowing through the body (amps)

(*iii*) *Duration* of the shocking current through the body,

Low voltage does not mean low hazard. Currents above 10 mA can paralyze or "freeze" muscles. Currents more than 75 mA can cause a rapid, ineffective heartbeat and death will occur in a few minutes unless a defibrillator is used.

Most common shock-related injury occurs when you touch electrical wiring or equipment that is improperly used or maintained and typically occurs on hands. (See Figure 1.1)

Electrical accidents are caused by a combination of three factors:

■ Unsafe equipment and/or installation,

■ Workplaces made unsafe by the environment, and

■ Unsafe work practices.

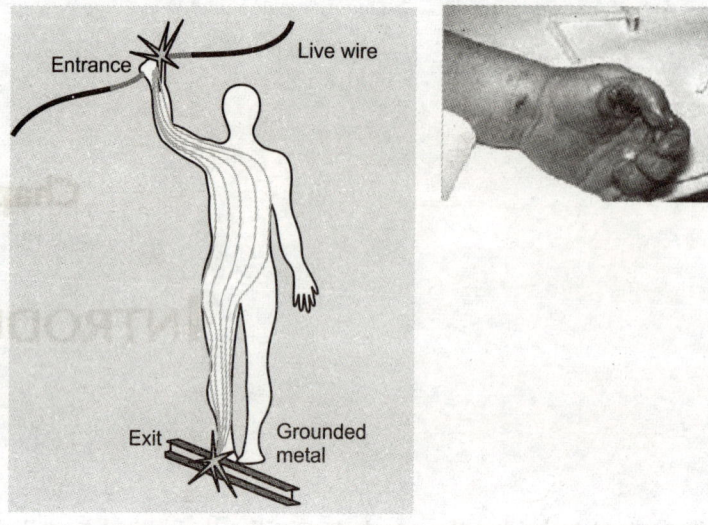

FIGURE 1.1

Electrical Hazard may also be due to wire too small in size for the current. The equipment or machine will draw more current than the wire can handle, causing overheating and a possible fire without tripping the circuit breaker or fuse. Hence, appropriate wire size must be used to avoid hazards. [*Wire gauge measures wires ranging in size from number 36 to 0 American wire gauge (AWG)*].

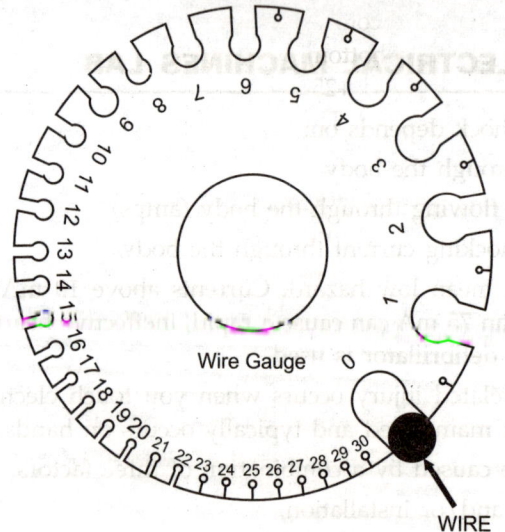

FIGURE 1.2. *Gauge meter*

The electrical engineering lab presents a potential hazardous environment in which to work. Care must be taken to prevent contact high voltage (415 V a.c. and 230 V d.c.) terminals and with rotating shafts. Every effort has been made to ensure that the equipment is as safe as possible while still allowing the flexibility to conduct experiments. A good attitude in the lab is to approach the equipment with caution and respect rather than fear.

FIGURE 1.3. *Control panel*

In the event you suspect that someone is in trouble, immediately switch off the main supply from the nearest control panel as shown in Figure 1.3 and alert the lab staff. Never touch a person whom you suspect of being in contact with electrical equipment until you are sure that the power supply is off.

In the event of arcing, short circuits, etc. stand clear. Do not attempt to disconnect the circuit, which may cause higher degree of panic. *Equipment,* though expensive, is *replaceable; you are not.*

In this lab we adhere to standard color code whenever possible; *i.e.* Three Phase a.c. is colored **(R)** Red, **(Y)** Yellow, **(B)** Blue, top to bottom or left to right. Neutrals are white and grounds are Green. Our d.c. system is colored Red +250V and Black 0, top to bottom or left to right. Strict adherence to the color code is expected. This makes it easy for you to wire circuits, and easy for the instructor to check connections.

The supply must be turned off while wiring.

The connection wires are provided in different lengths. A basic rule of good wiring is :

Never use a long lead where a shorter one will do. Always wire toward the supply. Connections to the contactor and starter should be made last.

POWER FROM THE MACHINE CONTROL PANNEL SHOULD **NOT BE TURNED ON** UNTIL YOUR CIRCUIT HAS BEEN CHECKED BY A LAB INSTRUCTOR, OR LAB TECHNICIAN. STUDENTS THAT DO NOT ADHERE TO THIS RULE MAY BE ASKED TO LEAVE THE LAB.

Position meters in such a way that they may be read without leaning over live connections.

1.2   BASIC EQUIPMENTS

The following basic equipments are used for lab experiments :

- **Single Phase Transformers :**

  400/230V, single phase, 50 Hz, 3kVA

  Center tapping on secondary side (230V),

  ±5%, ± 2.5%, tapping on primary side (400V)

FIGURE 1.4. *Single phase transformers*

■ **Three Phase Transformers :**

400/230, Δ-Y connected, 50 Hz, 5kVA

FIGURE 1.5. *Three phase transformers*

■ **Induction Motor (squirrel cage) with spring balance – for load test :**

3.7 kW/5 hp, 3-phase, 50 Hz, 415 V, 7.8 A, 1440 rpm, S1 – type

FIGURE 1.6. *Induction motor*

■ **d.c. shunt machine (motor or generator) :**

230 V dc, 5.5 hp, 1500 rpm, 21.7 Amp, coupled with

- **Synchronous machine (motor or generator) :**
  50 Hz, 3-phase, 415 V, 1500 rpm, 4kVA

FIGURE 1.7. *Synchronous machine*

- **DC shunt machine (Magnetization characteristics) :**
  230V dc, 4-pole, 1430 rpm, 12.6 A coupled with
- **Induction motor :**
  415 V ac, 50 Hz, 3-phase, 1430 rpm, 5 HP.

FIGURE 1.8. *Motor-generator set*

- **d.c. Compound Machine (Traction Motor) :**
  1000 rpm, 230 V dc, 20 A

FIGURE 1.9. *DC compound machine*

■ **d.c. Shunt machine (Hopkinson's Test) :**
230 V dc, 11 A, 1450 rpm.

FIGURE 1.10. *DC shunt machine*

■ **d.c. shunt machine with spring balance (Load Test) :**
230 V dc, 16 A, 1460 rpm, 5 hp.

FIGURE 1.11. *DC shunt machine with spring balance*

These machines are installed on the concrete foundation and their schematic and terminal diagrams are shown in connection diagram of different experiments.

- **Single Phase Auto-transformer :**

Input ac voltage 230 V, 50 Hz.

Output variable ac voltage 0–300 V.

FIGURE 1.12. *Single phase auto-transformer*

- **Three Phase Auto-transformer :**

Input voltage – 415 V, three-phase, 50 Hz, 5 kVA, Star connected.

Output Voltage 0-470 V

FIGURE 1.13. *Three phase auto-transformer*

Load Trolley : There are different type of load trolleys depending upon ratings and type of load like resistive, inductive or/and capacitive available in machines lab.

The load trolleys are shown in Figure 1.14. In resistive load trolley, there are many strips of bakelite former on which a nicrome wire is wound.

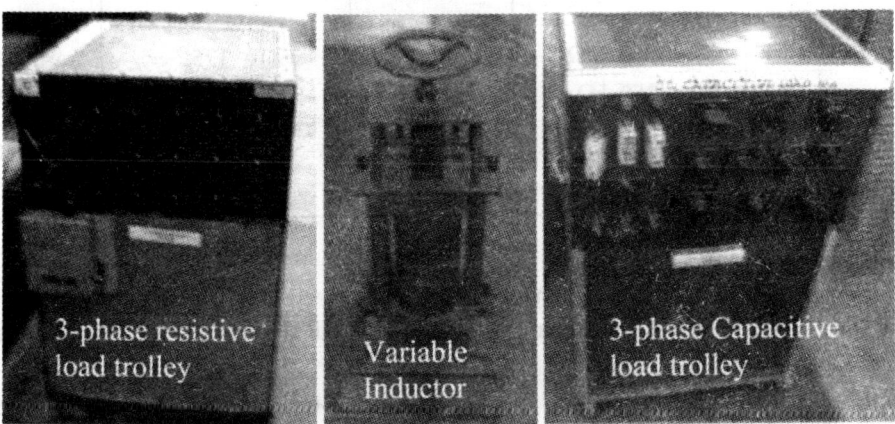

FIGURE 1.14. *Different types of load trolleys*

Resistors

Resistors can be broadly classified as fixed, variable, and special-purpose resistors. The characteristics of fixed resistors are shown in Table 1.1. The resistance for most resistors changes with temperature. The temperature coefficient of electrical resistance is the change in electrical resistance per unit change in temperature and measured in $\Omega/°C$. The temperature coefficient of resistors may be either positive or negative. A positive temperature coefficient denotes a rise in resistance with a rise in temperature; a negative temperature coefficient means a decrease in resistance with a rise in temperature. Pure metals typically have a positive temperature coefficient of resistance, while some metal alloys such as constantan and manganin have a zero temperature coefficient of resistance. Carbon and graphite mixed with binders usually exhibit negative temperature coefficients, although certain choices of binders and process variations may yield positive temperature coefficients. The temperature coefficient of resistance is given by

$$R_{T_2} = R_{T_1}[1 + \alpha(T_2 - T_1)]$$

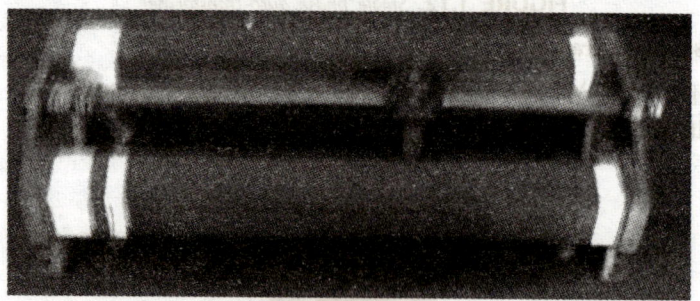

FIGURE 1.15. *Variable resistor*

TABLE 1.1. Characteristics of Typical Resistors

Resistor	Range	Watt	Operating Temp. range	α ppm/°C
Wire wound resistor				
Precision	0.1 to 1.2 MΩ	1/8 to 1/4	−55 to 145	10
Power	0.1 to 180 kΩ	1 to 210	−55 to 275	260
Metal film resistor				
Precision	1 to 250 MΩ	1/20 to 1	−55 to 125	50–100
Power	5 to 250 kΩ	1 to 5	−55 to 155	20–100
General purpose Composite resistor	2.7 to 100 MΩ	1/8 to 5	−55 to 130	1500

Capacitors

If a potential difference is found between two points, an electric **field** exists that is the result of the separation of unlike charges. The strength of the field will depend on the amount the charges have been separated.

Capacitance is the concept of energy storage in an electric field. This energy is stored and remains even after the input is removed. It depends on the area, shape, and spacing of the

capacitor plates and the property of the material separating them, which is called dielectric material.

The value of a parallel-plate capacitor can be found with the equation

$$C = \frac{x \in [(N-1)A]}{d} 10^{-13}$$

where

C = capacitance, F;

δ = dielectric constant of insulation;

d = spacing between plates;

N = number of plates;

A = area of plates; and

x = 0.0885 when A and d are in centimeters, and

x = 0.225 when A and d are in inches.

The **dielectric constant** of a material determines the electrostatic energy which may be stored in that material per unit volume for a given voltage. The dielectric of air is 1, and it is considered as the reference. As the dielectric constant is increased or decreased, the capacitance will increase or decrease, respectively. (Appendix–D, Table-2 shows the dielectric constants of various materials.) The dielectric constant of most of the materials is affected by both temperature and frequency, except for quartz, Styrofoam, and Teflon, whose dielectric constants remain essentially constant.

When a dc voltage is connected across a capacitor, a time t is required to charge the capacitor to the applied voltage. This is called a **time constant** and denoted by t.

$$t = RC$$

where

t = time, sec.;

R = resistance, Ω; and

C = capacitance, F.

In a circuit consisting of pure resistance and capacitance, the *time constant t* is defined as the time required to charge the capacitor to 63.2% of the applied voltage. The charge on a capacitor can never actually reach 100% but is considered to be 100% after five time constants. *Capacitance is expressed in microfarads (μF, or 10^{-6} F) or Pico farads (pF, or 10^{-12} F).*

Quality Factor (Q)

Quality factor is the ratio of the capacitor's **reactance** to its resistance at a specified frequency and is found by the equation

$$Q = \frac{1}{2\pi f CR} = \frac{1}{PF}$$

where

Q = quality factor;

f = frequency, Hz;

C = value of capacitance, F;

R = internal resistance, Ω; and

PF = power factor

Capacitors are used to filter, couple, tune, block dc, pass ac, bypass, shift phase, compensate, feed through, isolate, store energy, suppress noise, and start motors. Capacitors are grouped according to their dielectric material and mechanical configuration like :

- Ceramic capacitors,
- Film Capacitors,
- Mica capacitors,
- Paper-foil-filled capacitors and
- Electrolytic capacitors

Inductors

Inductance is used for the storage of magnetic energy. Magnetic energy is stored as long as current keeps flowing through the inductor. In a perfect **inductor**, the current of a sine wave lags the voltage by 90°. **Inductive reactance** X_L, the impedance of an inductor to an ac signal, is found by the equation

$$X_L = 2\pi f L$$

where
$$X_L = \text{inductive reactance, } \Omega;$$
$$f = \text{frequency, Hz; and}$$
$$L = \text{inductance, H.}$$

The type of wire used for its construction does not affect the inductance of a **coil**. Quality factor Q of the coil will be governed by the resistance of the wire. Therefore coils wound with silver or gold wire have the highest Q.

The total inductance will always be greater than the largest inductor.

$$L_T = L_1 + L_2 + ... + L_n$$

To reduce inductance, inductors are connected in parallel. The total inductance will always be less than the value of the lowest inductor.

1.3   MEASUREMENTS

1.3.1   Power Measurements

The instrument for measurement of average power flow is the wattmeter (WM). The wattmeter has a voltage or potential coil (PC) and a current coil (CC) with polarity marking as indicated. When the wattmeter connected with coil polarity marking, the meter simply reads the average power as

$$P = \frac{1}{T} \int_0^t v(t)dt$$

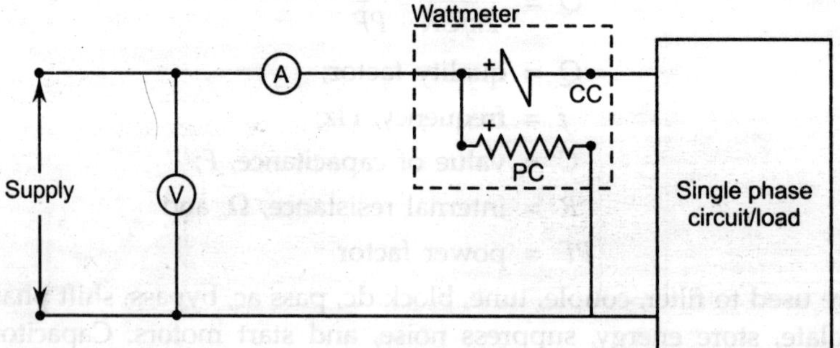

FIGURE 1.16. *Connection diagram for measurement of power, voltage and current*

If both coils are reversed, the wattmeter reading would be unchanged as shown in Figure 1.16. If either one of the coils are reversed, the meter will give the negative value of power. Digital meters will indicate any negative values by a minus sign in the display. Analog wattmeter will simply attempt to read down scale, and the user must reverse the potential coil connection and manually insert the negative value in the recorded data.

Measurement of average power flow in three phase circuits can be accomplished by connecting a wattmeter to read power flowing in each phase. The total three phase power would then be the sum of the three wattmeter readings. However, if the circuit is connected in star, the pressure coils of all three wattmeter be connected to the neutral. If neutral wire is not available (for example in three phase, three wire system) or load is delta connected then power measured by two wattmeter method as shown in Figure 1.17.

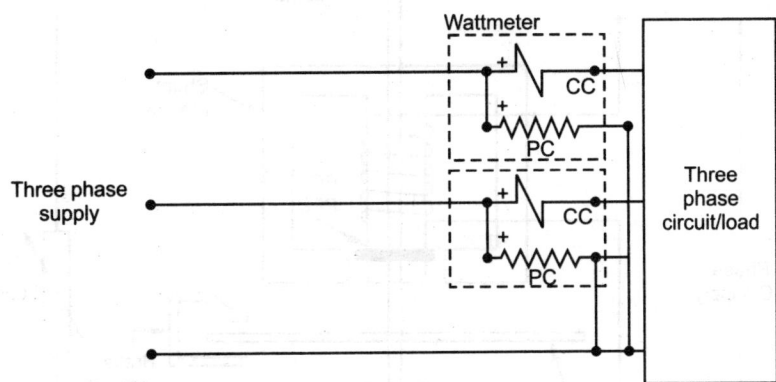

FIGURE 1.17. *Connection diagram for measurement of three phase power using two wattmeter method*

1.3.2 Energy Measurements

For the energy measurements mainly two types of energy meters are available as shown in Figure 1.18.

1. Static energy meters (Digital)
2. Rotating disc type energy meters (Analog/Digital).

FIGURE 1.18. *Static and rotating type energy meters*

The rotating disc type energy meter is also called induction type energy meter and consisting of the following systems as shown in Figure 1.19 :

1. Driving System
2. Moving System
3. Braking or Damping system and
4. Registering system.

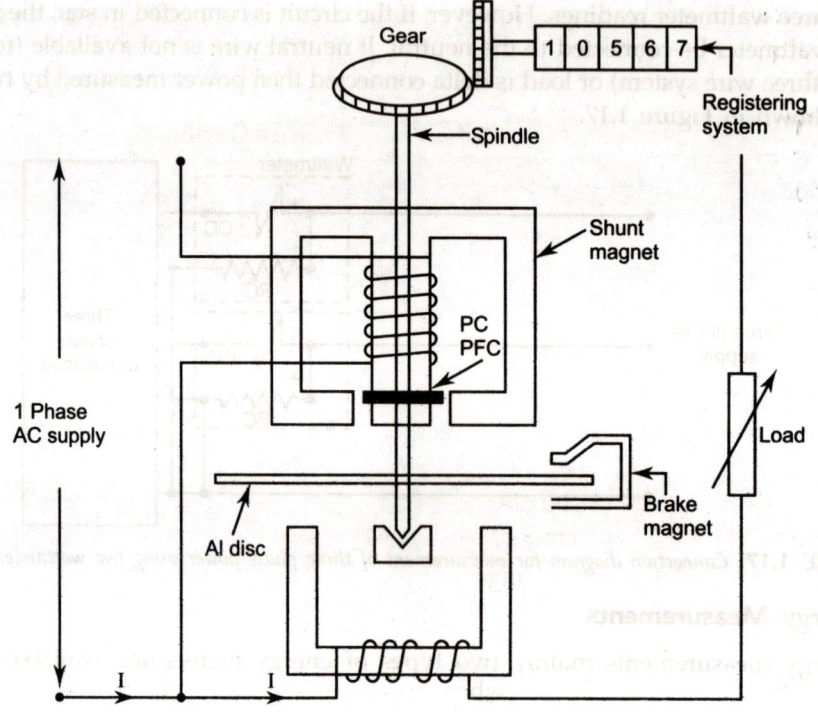

FIGURE 1.19. *Induction type rotating energy meter*

1.4 LAB NOTEBOOK

The standard hard cover lab book is required in the lab. The record for each experiment must start from the title page in which a title, date of the experiment and date of submission, as well as experiment number shall appear. This record shall consist of two parts:

Part I : Reading taken in the lab, *i.e.,* rough records on the experiment shall be organised in the observation table and, if possible, a graph is to be drawn. This part must be checked and signed by the lab staff just after performing the experiment;

Part II : Experiment write up is to be prepared and submitted later on along with part I.

Based on the experimental results, all required characteristics, phasor diagrams etc. must be presented with a short and concise explanation with regards to their meaning and/or interpretation. Lab notebook record consists of the following information for each experiment:

(*i*) Date of experiment/Date of submission

(*ii*) Experiment number

(*iii*) Title of the experiment

(iv) Need of the experiment

(v) List of instruments and devices required along with their specifications.

(vi) Procedural steps

(vii) Observations

(viii) Calculations and plots

(ix) Conclusions

(x) Precautions (if any).

The sample format of experiment is given in the appendix A.

1.5 CHARACTERISTICS OF ELECTRICAL MACHINES

1.5.1 AC Motors (Three Phase)

(a) **Induction motor :**

Power range 1–5000 hp.

Stator : 3-phase armature winding

Rotor : Squirrel cage

Advantage. Simple rugged construction. Commonly used in fan, blowers, pumps. If Rotor has wound field then variable adjustable speed using rotor resistance may be obtained and it is used for applications like cranes, hoists, etc.

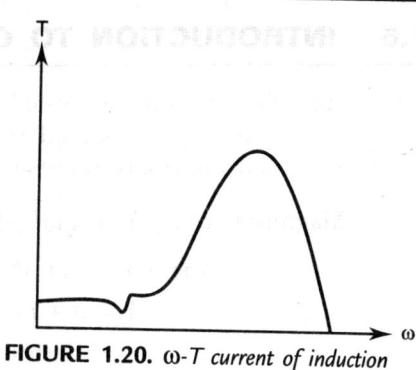

FIGURE 1.20. ω-T current of induction motor

(b) **Synchronous motor :** Power range 1–5 hp and rotor is permanent magnet.

Application. Used in

(i) Transport of sheet material where precise speed is critical.

(ii) Electric clocks

Power range : 1,000–50,000 hp

Rotor : DC field winding.

Used for driving large constant loads and for power factor correction.

1.5.2 DC Motors

(a) **Shunt motors :**

Application. Used in precise and smooth speed control in a wide range like Rolling mills, paper mills, textile mills etc.

Rotor contains armature winding, and stator has field and Inter pole windings.

It has fairly constant speed. Speed slightly reduces as load increases. This is called drooping characteristic of dc shunt motors.

(b) **Series motors :**

Application. used in mixer machines, lifts and cranes.

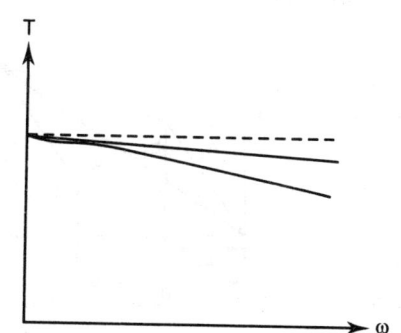

FIGURE 1.21. ω-T characteristic of DC shunt motor

The speed–torque characteristics of dc series motors is self releasing means as load increases speed reduces and if the load torque reduces speed increases.

(c) **Compound motors :**

Applications. The traction motors in electric locomotives are compound machines whose characteristic are close to dc series motors. In these machines torque can easily be controlled to any value at the time of starting as well as running conditions.

In ac circuits all the quantities are in complex form and it is necessary to study the complex numbers in detail.

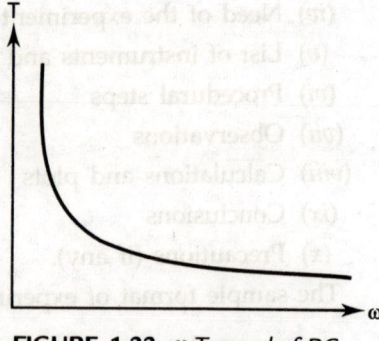

FIGURE 1.22. ω-T curved of DC series motor

1.6   INTRODUCTION TO COMPLEX NUMBERS

Argand in 1806, first time suggested the complex number in the form of $A = a + jb$, could be represented graphically by plotting real number 'a' on x-axis and imaginary number 'b' on y-axis.

$$\text{Magnitude of } A,\ A = \sqrt{(a^2 + b^2)}$$

$$\text{Angle } \theta = \tan^{-1}(b/a)$$

$$A = a + j\,b \text{ is rectangular or}$$

FIGURE 1.23. *Representation of complex number $A = a + jb = \sqrt{(a^2 + b^2)}\,\angle\theta$*

Cartesian notation

$$= A(\cos\theta + j\sin\theta) \text{ is trigonometric notation}$$

$$= A\,\angle\theta \text{ is polar notation.}$$

If there are two complex numbers $A_1 = a_1 + j\,b_1$ and $A_2 = a_2 + j\,b_2$

Then addition of these two complex numbers will be

$$A_1 + A_2 = (a_1 + a_2) + j(b_1 + b_2)$$

and the subtraction of these two complex numbers will be

$$A_1 - A_2 = (a_1 - a_2) + j(b_1 - b_2).$$

Hence, the addition and subtraction of any two complex numbers are easy if the numbers are in Cartesian form.

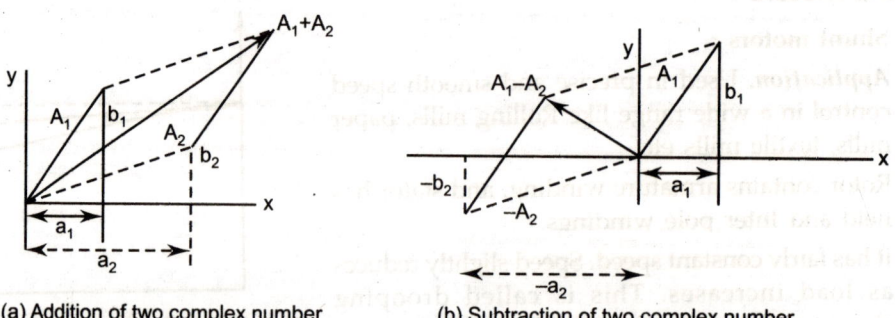

(a) Addition of two complex number   (b) Subtraction of two complex number

FIGURE 1.24. *Addition and subtraction of two complex numbers*

Similarly, the product and division of any two complex numbers will be easy in polar form, *e.g.*,

$$\text{Product } A_1{}^*A_2 = A_1 \angle\theta * A_2 \angle\theta_2 = (A_1 * A_2) \angle(\theta_1 + \theta_2) \text{ and}$$
$$\text{Division of } A_1/A_2 = A_1 \angle\theta_1/A_2 \angle\theta_2 = (A_1/A_2) \angle(\theta_1 - \theta_2)$$

PROBLEM 1. *Write a matlab program to*

(i) *change the form of complex number from Cartesian to polar form*

(ii) *multiply two or more complex numbers*

The Matlab program changes the form from polar to Cartesian and vice versa, if there is only one complex number. If there are n–complex numbers, the program will multiply them. Where n is greater or equal to two.

For example, Complex Number 1 = 10 $\angle 45°$ and

Complex Number 2 = 5 + $j*2$

When these numbers are multiplied, the product will be

$$= 21.2132 + j\,49.4975 = 53.8516 \angle 66.8014 \text{ deg}$$

MATLAB PROGRAM-1

```
%***********************************************************
% complex_prod.m -
% i. Change the form of complex number from polar and rectangular and vice
     versa.
% ii. Calculates the product of complex numbers
%***********************************************************
clear all
disp('Change the form of complex number')
disp('or')
disp('Multiplication of two or more complex numbers');
disp('')
disp('How many numbers to be multiplied?')
disp('(if answer is 1 then the program change the form of complex number)');
n=input('Give number of complex variables ');
disp(' ')
for i=1:n;
    disp('Form of complex number ');
    disp('1-polar, 2-rectangular');
    F=input(['Give your choice for form of complex number ',num2str(i),' = ']);
    if (F~=1 & F~=2); disp('INVALID FORM'); end
    if F==1
        M=input(['Mag ',num2str(i),' = ']);
        A=input(['Deg ',num2str(i),' = '])*pi/180;
        N(i)=M*exp(j*A); disp(' ');
    else
        R=input(['Real ',num2str(i),' = ']);
        I=input(['Imag ',num2str(i),' = ']);
        N(i)=R+j*I; disp(' ');
    end
end
P=N(1); for k=2:n; P=P*N(k); end
disp(' '); disp(['Answer = ' num2str(real(P)) ' +j '...
```

```
        num2str(imag(P)) ' = ' num2str(abs(P)) '|_'...
        num2str(angle(P)*180/pi) 'deg']);
%******************************************************
```

PROBLEM 2. *Express the rectangular and polar form of impedance for the following circuit at 50Hz frequency:*

(i) *A resistance R = 15 Ω in series with an inductance of 0.1 H.*

(ii) *Also calculate the voltage if current in the circuit is (1 + j2) A.*

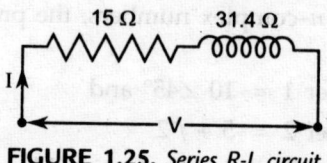

FIGURE 1.25. *Series R-L circuit*

(i) For 50 Hz

$$\omega = 2\pi f = 2\pi \times 50 = 314 \text{ rad/s}$$
$$Z = 15 + j\,314 \times 0.1 = (15 + j\,31.4)\ \Omega$$

If this information is passed on to the matlab program, following results will be obtained.

PROGRAM OUTPUT

Change the form of complex number

Or

Multiplication of two or more complex numbers

How many numbers to be multiplied?

(If answer is 1 then the program change the form of complex number)

Give number of complex variables 1

Form of complex number

1-polar, 2-rectangular

Give your choice for form of complex number 1 = 2

Real 1 = 15

Imag 1 = 31.4

Answer = 15 + j 31.4 = 34.7989 ∠64.4659deg

(ii) The voltage $V = I \times Z = (1 + j\,2) \times (15 + j\,31.4)$

PROGRAM OUTPUT

Change the form of complex number

Or

Multiplication of two or more complex numbers

How many numbers to be multiplied?

(If answer is 1 then the program change the form of complex number)

Give number of complex variables 2

Form of complex number

1-polar, 2-rectangular

Give your choice for form of complex number 1 = 2
Real 1 = 1
Imag 1 = 2
Form of complex number
1-polar, 2-rectangular
Give your choice for form of complex number 2 = 2
Real 2 = 15
Imag 2 = 31.4
Answer = –47.8 + j 61.4 = 77.8126 ∠127.9008deg

PROBLEM 3. *Write a matlab program for finding the ratio of two complex numbers.*

```
%****************************************************************
% complex_ratio.m - Calculates the ratio of complex numbers
%****************************************************************
disp('');
disp('RATIO OF COMPLEX NUMBER PRODUCTS'); disp('');
disp('Form: 1-polar, 2-rectangular'); disp('');
num=input('How many numerator numbers? '); disp(' ');
for i=1:num;
    F=input(['Form of ',num2str(i),' = ']);
    if (F~=1 & F~=2); disp('INVALID FORM'); end
    if F==1
        M=input(['Mag ',num2str(i),' = ']);
        A=input(['Deg ',num2str(i),' = '])*pi/180;
        N(i)=M*exp(j*A); disp(' ');
    else
        R=input(['Real ',num2str(i),' = ']);
        I=input(['Imag ',num2str(i),' = ']);
        N(i)=R+j*I; disp(' ');
    end
end
NP=N(1); for k=2:num; NP=NP*N(k); end
den=input('How many denominator numbers? '); disp(' ');
for i=1:den;
    F=input(['Form of ',num2str(i),' = ']);
    if (F~=1 & F~=2); disp('INVALID FORM'); end
    if F==1
        M=input(['Mag ',num2str(i),' = ']);
        A=input(['Deg ',num2str(i),' = '])*pi/180;
        D(i)=M*exp(j*A); disp(' ');
    else
        R=input(['Real ',num2str(i),' = ']);
        I=input(['Imag ',num2str(i),' = ']);
        D(i)=R+j*I; disp(' ');
    end
end
```

```
DP=D(1); for k=2:den; DP=DP*D(k); end; RAT=NP/DP;
disp(' '); disp(['RATIO = ' num2str(real(RAT)) ' +j '...
    num2str(imag(RAT)) ' = ' num2str(abs(RAT)) '|_'...
    num2str(angle(RAT)*180/pi) 'deg']);
%************************************************************
```

PROBLEM 4. *Find the current through an impedance Z = (10 + j 13.5) ohm, if terminal voltage of 230V, 50 Hz is applied.*

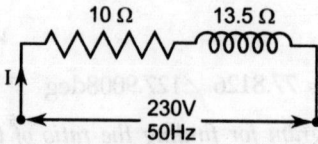

FIGURE 1.26. *Series R-L circuit*

With the help of above matlab program we can easily find the current as

$$I = \frac{V}{Z} = \frac{230 + j0}{10 + j13.5}$$

PROGRAM OUTPUT

RATIO OF COMPLEX NUMBER PRODUCTS

Form: 1-polar, 2-rectangular

How many numerator numbers? 1

Form of 1 = 2

Real 1 = 230

Imag 1 = 0

How many denominator numbers? 1

Form of 1 = 2

Real 1 = 10

Imag 1 = 13.5

RATIO = 8.1488 +j -11.0009 = 13.6902 |_-53.4711deg.

PROBLEM 5. *Write a Matlab program for adding two or more than two complex numbers in polar or Cartesian form.*

```
%************************************************************
%
% cumulative_sum.m - forms the sum of a series of complex numbers
%
%************************************************************
clear all
S=0; disp(' ');
disp(' SUM OF COMPLEX NUMBERS'); disp(' ');
disp(' Form: 1-polar, 2-rectangular'); disp(' ');
nterms=input('How many numbers to be added? '); disp(' ');
for i=1:nterms;
    F=input(['Form of ',num2str(i),' = ']);
    if F==1
        M=input(['Mag ',num2str(i),' = ']);
```

```
        A=input(['Deg ',num2str(i),' = '])*pi/180;
        R=M*cos(A);
        I=M*sin(A);
        S=S+R+j*I; disp(' ');
    else
        R=input(['Real ',num2str(i),' = ']);
        I=input(['Imag ',num2str(i),' = ']);
        S=S+R+j*I; disp(' ');
    end
end
disp(' '); disp(['SUM = ' num2str(real(S)) ' +j '...
    num2str(imag(S)) ' = ' num2str(abs(S)) '|_'...
    num2str(angle(S)*180/pi) 'deg']);
%************************************************************
```

PROBLEM 6. *Write a matlab program to determine the total impedance of the circuit shown in Figure 1.27.*

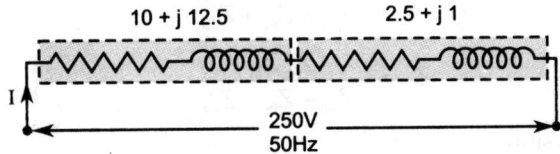

10 + j 12.5   2.5 + j 1

I

250V
50Hz

FIGURE 1.27. *Series circuit consisting of two impedances*

Program output of Cumulative_sum.m

SUM OF COMPLEX NUMBERS

Form: 1-polar, 2-rectangular

How many numbers to be added? 2

Form of 1 = 2

Real 1 = 10

Imag 1 = 12.5

Form of 2 = 2

Real 2 = 2.5

Imag 2 = 1

SUM = 12.5 + j 13.5 = 18.3984 ∠47.2026deg

PROBLEM 7. *Write a Matlab program for converting delta connected load into equivalent star connected load and vice versa for balanced or unbalanced load when load impedances may be in polar form or cartesian form.*

In a three phase system the 3-phase alternators are connected to 3-phase loads. There are two common topological arrangements (such as star and delta) for connection of the source and load. In star connection, the circuit element is connected between one line and the common neutral as shown in Figure 1.28(*a*).

In case of delta connected loads the elements connected between two lines such as connected between lines *a-b*, lines *b-c*, lines *c-a* as shown in Figure 1.28(*b*)

In balanced conditions, the voltages and currents in each phase of a 3-phase system are equal in magnitude and phase angles are 120° apart. Hence, in balanced star connected load, the voltage may be written as:

$$V_{an} = V_m \angle \varphi$$
$$V_{bn} = V_m \angle (\varphi + 120),$$
$$V_{cn} = V_m \angle (\varphi + 240).$$

Conversion

(i) delta-star conversion

The star connected impedances may be calculated from the given delta connected impedances.

$$Z_a = \frac{Z_{ab}Z_{ca}}{Z_{ab} + Z_{bc} + Z_{ca}}$$

$$Z_b = \frac{Z_{bc}Z_{ab}}{Z_{ab} + Z_{bc} + Z_{ca}}$$

$$Z_c = \frac{Z_{ca}Z_{bc}}{Z_{ab} + Z_{bc} + Z_{ca}}$$

(ii) star-delta conversion

The delta connected impedances may be calculated from the given star connected impedances.

$$Z_{ab} = \frac{Z_aZ_b + Z_bZ_c + Z_cZ_a}{Z_c}$$

$$Z_{bc} = \frac{Z_aZ_b + Z_bZ_c + Z_cZ_a}{Z_a}$$

$$Z_{ca} = \frac{Z_aZ_b + Z_bZ_c + Z_cZ_a}{Z_b}$$

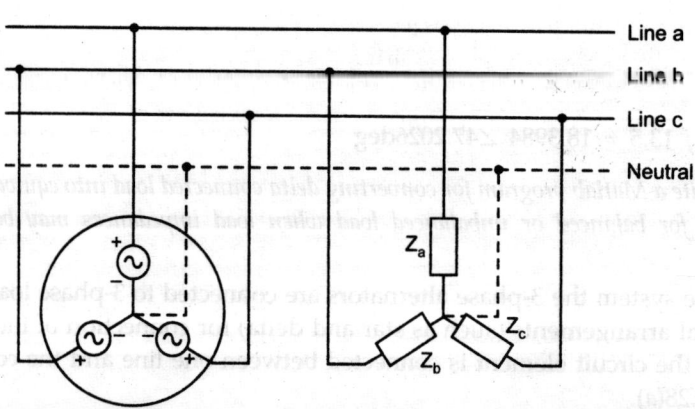

Line a

Line b

Line c

Neutral

Z_a

Z_c

Z_b

FIGURE 1.28. (a) AC alternator and star connected load

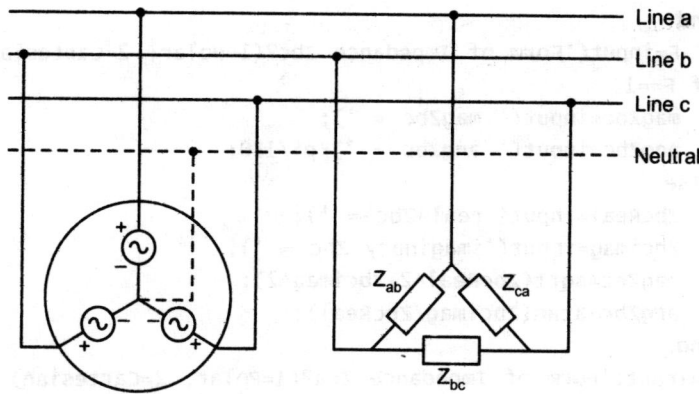

FIGURE 1.28. (*b*) *Star connected AC alternator and delta connected load*

```
%*****************************************************************
%
% delta_star.m - performs delta-star or star-delta conversion
%
%*****************************************************************
clear all;
Dir=input(' Conversion From? (1=Delta->Star, 2=Star->Delta) '); disp(' ');
Bload=input(' Balanced Load? (1=Yes, 2=No) '); disp(' ');
if Dir==1
    disp(' DELTA to STAR CONVERSION'); disp(' ');
    if Bload==1
        F=input('Form of Impedance Zab?(1=Polar, 2=Cartesian) ');
        if F==1
            magZab=input(' magZab = ');
            angZab=input(' angZab = ')*pi/180;
        else
            ZabReal=input('real Zab = ');
            Zabimag=input('imaginary Zab = ');
            magZab=sqrt(ZabReal^2+Zabimag^2);
            angZab=atan(Zabimag/ZabReal);
        end
        magZbc=magZab; magZca=magZab;
        angZbc=angZab; angZca=angZab;
    elseif Bload==2
        F=input('Form of Impedance Zab?(1=Polar, 2=Cartesian)');
        if F==1
            magZab=input(' magZab = ');
            angZab=input(' angZab = ')*pi/180;
        else
            ZabReal=input('real Zab = ');
            Zabimag=input('imaginary Zab = ');
            magZab=sqrt(ZabReal^2+Zabimag^2);
            angZab=atan(Zabimag/ZabReal);
```

```
          end
            F=input('Form of Impedance Zbc?(1=Polar, 2=Cartesian) ');
          if F==1
            magZbc=input(' magZbc = ');
            angZbc=input(' angZbc = ')*pi/180;
          else
            ZbcReal=input('real Zbc = ');
            Zbcimag=input('imaginary Zbc = ');
            magZbc=sqrt(ZbcReal^2+Zbcimag^2);
            angZbc=atan(Zbcimag/ZbcReal);
          end
          F=input('Form of Impedance Zca?(1=Polar, 2=Cartesian) ');
          if F==1
            magZca=input(' magZca = ');
            angZca=input(' angZca = ')*pi/180;
          else
            ZcaReal=input('real Zca = ');
            Zcaimag=input('imaginary Zca = ');
            magZca=sqrt(ZcaReal^2+Zcaimag^2);
            angZca=atan(Zcaimag/ZcaReal);
          end
        end
        Zab=magZab*exp(j*angZab); Zbc=magZbc*exp(j*angZbc);
        Zca=magZca*exp(j*angZbc); den=Zab+Zbc+Zca;
        Za=Zab*Zca/den; Zb=Zbc*Zab/den; Zc=Zca*Zbc/den;
        disp(' '); disp(' Star connected Impedances = ');
        disp(' Mag Angle');
        disp([abs(Za) angle(Za)*180/pi; abs(Zb) angle(Zb)*180/pi; ...
            abs(Zc) angle(Zc)*180/pi]);
    else if Dir==2
      disp(' STAR to DELTA CONVERSION'); disp(' ');
      if Bload==1
        F=input('Form of Impedance Za?(1=Polar, 2=Cartesian) ');
        if F==1
          magZa=input(' magZa = ');
          angZa=input(' angZa = ')*pi/180;
    else
        ZaReal=input('real Za = ');
        Zaimag=input('imaginary Za = ');
        magZa=sqrt(ZaReal^2+Zaimag^2);
        angZa=atan(Zaimag/ZaReal);
    end
        magZb=magZa; magZc=magZa;
        angZb=angZa; angZc=angZa;
    else if Bload==2
      F=input('Form of Impedance Za?(1=Polar, 2=Cartesian) ');
      if F==1
        magZa=input(' magZa = ');
```

```
        angZa=input(' angZa = ')*pi/180;
    else
        ZaReal=input('real Za = ');
        Zaimag=input('imaginary Za = ');
        magZa=sqrt(ZaReal^2+Zaimag^2);
        angZa=atan(Zaimag/ZaReal);
    end
  F=input('Form of Impedance Zb?(1=Polar, 2=Cartesian) ');
  if F==1
     magZb=input(' magZb = ');
     angZb=input(' angZb = ')*pi/180;
  else
     ZbReal=input('real Zb = ');
     Zbimag=input('imaginary Zb = ');
     magZb=sqrt(ZbReal^2+Zbimag^2);
     angZb=atan(Zbimag/ZbReal);
  end
  F=input('Form of Impedance Zc?(1=Polar, 2=Cartesian) ');
  if F==1
     magZc=input(' magZc = ');
     angZc=input(' angZc = ')*pi/180;
  else
     ZcReal=input('real Zc = ');
     Zcimag=input('imaginary Zc = ');
     magZc=sqrt(ZcReal^2+Zcimag^2);
     angZc=atan(Zcimag/ZcReal);
     end
  end
  Za=magZa*exp(j*angZa);  Zb=magZb*exp(j*angZb);
  Zc=magZc*exp(j*angZc);  num=Za*Zb+Zb*Zc+Zc*Za;
  Zab=num/Zc;  Zbc=num/Za;  Zca=num/Zb;
  disp(' ');
  disp(' DELTA IMPEDANCES in POLAR FORM');
  disp(' Mag Angle');
  disp([abs(Zab) angle(Zab)*180/pi; ...
     abs(zbc) angle(zbc)*180/pi; ...
     abs(Zca) angle(Zca)*180/pi]);
  end
  %**************************************************************
```

PROBLEM 8. *Let us consider star connected impedances*

$$Z_a = 3 + j * 4$$
$$Z_b = 10 \angle 30°$$
$$Z_c = 3 \angle -45°$$

Find the equivalent delta connected impedances.

PROGRAM OUTPUT

Conversion From? (1=Delta->Star, 2=Star->Delta) 2

Balanced Load? (1=Yes, 2=No) 2

STAR to DELTA CONVERSION

Form of Impedance Za?(1=Polar, 2=Cartesian) 2

real Za = 3

imaginary Za = 4

Form of Impedance Zb?(1=Polar, 2=Cartesian) 1

  magZb = 10

  angZb = 30

Form of Impedance Zc?(1=Polar, 2=Cartesian) 1

  magZc = 3

  angZc = –45

DELTA IMPEDANCES in POLAR FORM

Mag	Angle
22.1525	86.4558
13.2915	–11.6743
6.6458	11.4558

PROBLEM 9. *Let us consider delta connected impedances*

$$Z_{ab} = 22 + j *6$$
$$Z_{bc} = 15 \angle -24.5°$$
$$Z_{ca} = 7 \angle 11°$$

Find the equivalent star connected impedances.

PROGRAM OUTPUT

Conversion From? (1=Delta->Star, 2=Star->Delta) 1

Balanced Load? (1=Yes, 2=No) 2

DELTA to STAR CONVERSION

Form of Impedance Zab?(1=Polar, 2=Cartesian) 2

  real Zab = 22

  imaginary Zab = 6

Form of Impedance Zbc?(1=Polar, 2=Cartesian) 1

  magZbc = 15

  angZbc = -24.5

Form of Impedance Zca?(1=Polar, 2=Cartesian) 1

  magZca = 7

  angZca = 11

Star connected Impedances =

Mag	Angle
3.7884	–4.9939
8.1180	–4.9939
2.4920	–44.7491

Important Note for Matlab Program

1. Blank lines and blank spaces not important, but blank space in the square bracket separates the two columns.
2. Semicolons are important.

   [1 2; 2 3]—semicolon separates the two rows

       $A = B;$—semicolon suppress the results (not to show on screen)
3. Anything to right of % is a comment not executed.
4. MATLAB is case sensitive, that is B and b are different.
5. Watch out for 1 (one) and l (lower case L).

Chapter 2

EXPERIMENTS FOR MACHINES LAB

SECTION-A

2.1 TRANSFORMER

One of the most significant static electrical devices is the transformer which contains magnetically coupled windings and takes in power at one voltage level and sends it out at another voltage level at same frequency. This conversion helps in the transmission of electrical power, since at higher voltages the lines carry low currents hence incur low losses.

2.1.1 Introduction

One of the main advantages of a.c. transmission and distribution is the ease with which an alternating voltage can be increased or reduced. For instance, the general practice is to generate electrical power at 11 kV, 33 kV, or 66 kV and then step up by means of transformers to higher voltages for transmission purpose. Near the load center, the voltage of electrical power is stepped down to values suitable to operate different loads like motors, lamps, heaters, etc. Usually transformers have a full load efficiency of about 97–98%, so that the loss at each point of transformation is small (about 2–3%, which is not significant). The high efficiency in case of transformers is due to non movable part in it. Hence, the amount of supervision is practically negligible.

2.1.2 Classification of Transformer

The major types of transformers for both power and communication applications are listed below, along with general operating frequencies for each type.

Power :

Power/distribution transformer	50 or 60 Hz
Instrument transformer	50 or 60 Hz
Ferro resonant transformer	50 or 60 Hz
Converter transformer	100 to 150 Hz.

Communication :

Audio–frequency transformer	20 Hz to 20 kHz
Carrier–frequency transformer	20 kHz to 20 MHz
High frequency transformer	20 MHz to 1000 MHz
Pulse transformer	Repetition rates to 4MHz

The transformers used with power system applications called power or distribution transformers. Power transformers are generally used near generating stations and always operate at full load. Hence, they designed for maximum efficiency near full load. Contrary to power transformers, distribution transformers operate at light loads most of the time and therefore they are designed to get maximum efficiency at light/average loads. Transformers are also used in many low power applications including electronic/communication circuits.

Instrument Transformers : Some of the transformers are used to reduce voltage or current to measure with normal measuring devices called instrument transformers. There are two types of instrument transforms.

(i) *Potential Transformer (PT)* : This is a transformer with a special design and construction. It is often referred as PT. It is used to step down high voltages so that they fall within the range of a voltmeter. A typical connection diagram is shown in Figure 2.1.

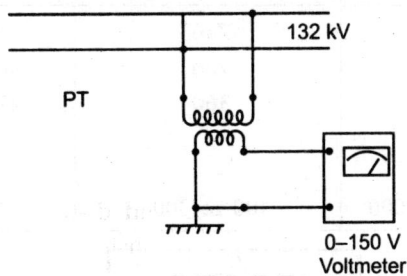

FIGURE 2.1. *Connection diagram for PT*

(ii) *Current Transformer (CT)* : This is also a transformer with a special design and construction. A current transformer often referred as CT. It is used to step down large currents so that they fall within the range of an ammeter. A typical connection diagram is shown in Figure 2.2.

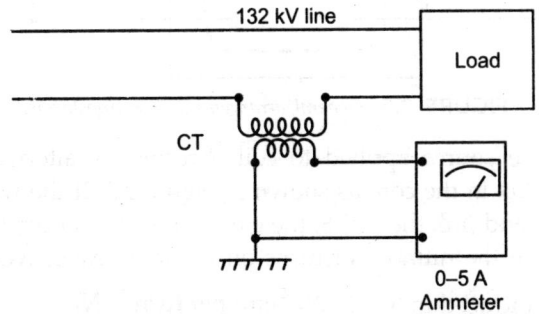

FIGURE 2.2. *Connection diagram for CT*

 (i) Always keep the secondary of CT short circuited, otherwise under open circuited conditions dangerously high voltage will appear on the secondary terminals, which may cause accidents.

(ii) Similarly, always keep the secondary of PT open circuited. Under short circuited secondary large amount of current will flow and may damage PT.

2.1.3 Principle of Operation

The general arrangement of transformer is shown in Figure 2.3. There are two windings wound on a laminated silicon steel core of about 0.35–0.7 mm thick laminations, insulated from one another. The vertical portion of the core is called **limbs** and the top and bottom portions are referred as **yokes**. The type of core material is known as soft magnetic material, which is defined as ferromagnetic material. It can be easily magnetized and demagnetized with a little coercive force (MMF). These materials have relatively high value of permeability and low hysterisis loss. The principal ferromagnetic materials are iron, nickel, cobalt and certain other alloys. The properties of soft magnetic materials are given in Table 2.1.

TABLE 2.1. The Properties of Soft Magnetic Materials

Materials	Permeability	Currie temperature (°C)	Resistivity	Operating frequency (Hz)
SiFe (unoriented)	400	740	47×10^{-6}	50 to 1000
SiFe (oriented)	1500	740	50×10^{-6}	50 to 1000
NiFe(oriented) Ni-50% Fe-50%	2000	360	47×10^{-6}	50 to 1000
Ferrite-MnZn	750 to 15000	100 to 300	10 to 100	10 k to2 M

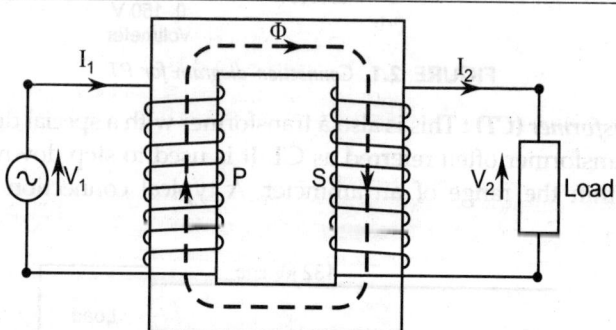

FIGURE 2.3. *General arrangement of a transformer*

An alternating voltage source applied to coil P results in alternating current flow, which produces an alternating flux in the core as shown in Figure 2.3. If the whole of the flux produced by P passes through core and link the coil S, the e.m.f. induced in each turn is the same for P and S. Hence, if N_1 and N_2 are the number of turns on P and S respectively.

$$\frac{\text{Total emf induced in } S}{\text{Total emf induced in } P} = \frac{N_1 * \text{emf per turn}}{N_2 * \text{emf per turn}} = \frac{N_1}{N_2}$$

When the secondary is open circuit, its terminal voltage is the same as the induced emf. The primary current is then very small, so that applied voltage V_1 is practically equal to the emf induced in P.

Hence $$\frac{V_1}{V_2} \cong \frac{N_1}{N_2}$$

Since the full load efficiency of transformer is nearly 100%, input power P_1 is same as output power P_2.

$$P_1 = P_2 \rightarrow V_1\, I_1{}^* \text{ power factor of coil } P = V_2\, I_2{}^* \text{ power factor of coil } S$$

The power factors at full load are nearly equal at both the coils.

$$\frac{V_1}{V_2} \cong \frac{I_2}{I_1} \cong \frac{N_1}{N_2}$$

Hence, $N_1\, I_1 = N_2\, I_2 \rightarrow$ ampere turns of coil $P =$ ampere turns of coil S

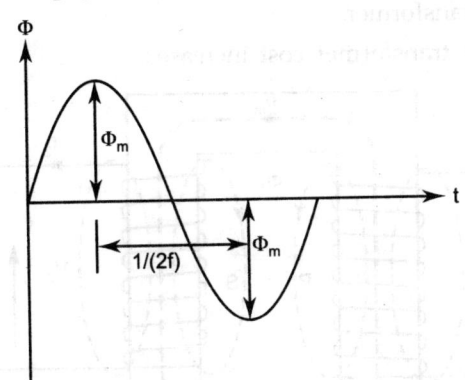

FIGURE 2.4. *Waveform of flux variation*

Let the maximum flux be Φ_m and the applied frequency be f hertz. Then the flux changes from $+ \Phi_m$ to $- \Phi_m$ in $1/(2f)$ seconds.

Average rate of change of flux $= \dfrac{2\Phi_m}{1/(2f)} = 4\,\Phi_m f$ Webers/second and the average induced emf per turn $= 4\,\Phi_m f$ volts. But in sinusoidal wave the rms or effective value is 1.11 times the average value.

The rms value of induced emf per turn $= 1.11 * 4\,\Phi_m f = 4.44\,\Phi_m f$ volts

Hence, the total rms value of emf induced in coil P, $E_1 = 4.44\,\Phi_m f\, N_1$ volts, and the total rms value of emf induced in coil S, $E_2 = 4.44\,\Phi_m f\, N_2$ volts.

Concepts of Leakage Flux

The flow of current in coil S due to load sets up a flux in the opposite direction to the main flux by which it is produced as shown in Figure 2.5. This flux returns through air. Since the flux linked only with the windings by which it is produced, referred as leakage flux and is responsible for inducing an emf of self inductance. The reluctance of the path of leakage flux is entirely due to the long air path and is therefore practically constant. Consequently the value of leakage flux is proportional to the load current, whereas the value of useful flux remains constant and almost independent of load current. The value of leakage flux is relatively small due to high reluctance path of air.

From the above discussion it is clear that the actual flux in transformer can be regarded as being due to two components namely,

(*i*) The useful flux, which links both the coils, is practically constant.

(*ii*) The leakage flux links only the coil by which it is produced and its value depends on the load.

When the transformer operates at no load the current drawn by the coil P is mainly used to keep the steel core magnetized and the transformer acts as a choke coil, having very high inductance and very low resistance. Hence, the power factor of transformer at no load is very low. To reduce or eliminate the leakage flux both coils P and S would on the same limb one over the other. There are some practical methods to reduce or eliminate leakage flux:

(*i*) Making the transformer window long and narrow.

(*ii*) Arranging the coils P and S concentrically.

(*iii*) Sandwiching the coils P and S.

(*iv*) Use of shell type transformer.

In all these methods the transformer cost increases.

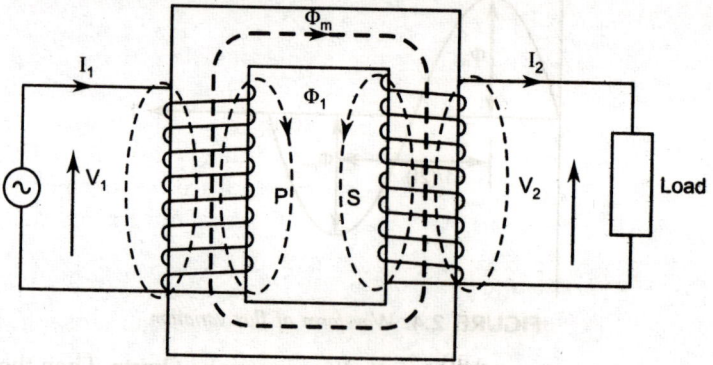

FIGURE 2.5. *Path of leakage flux in a transformer*

Power Transformer

The power transformers as used in a power station or a power system could be a bank of three single phase transformers connected in either star/delta or star/star etc., or could be a single 3-phase transformer unit with single core as shown in Figure 2.6. Normally for large capacity transformers, a three-phase transformer is used for the following reasons. It is lighter and cheaper, occupies less space and is more efficient. The only disadvantage is that anything that affects the winding of one phase will affect the others also whereas in single phase transformers this is not so, as one transformer can be replaced and the operation can be continued.

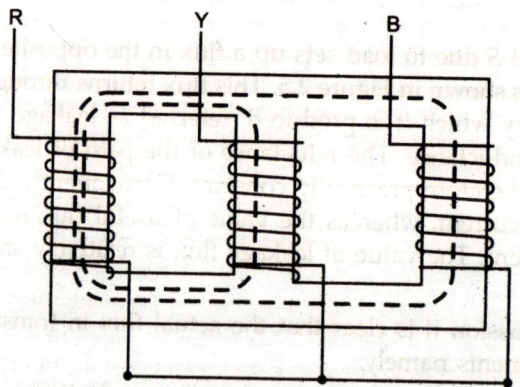

FIGURE 2.6. *Three phase core type transformer*

For all practical purposes the flux is same in all parts of the magnetic circuit and the cross-section of the yoke and the limbs should be same in order to have uniform flux density everywhere.

The type of connections for 3-phase operation of transformers normally used is delta-star, star-delta and delta-delta. Star-Star connections are normally not considered. The delta star and star-delta connections are well suited to transformers in high voltage systems, the delta-star connection being used for stepping up the voltage and the star-delta for stepping down. Star connection is used for high voltages as the phase voltage is equal to the line voltage divided by $\sqrt{3}$ and thus the windings can be insulated for lower voltages thereby the cost of providing insulation is reduced. The delta winding is used as low voltage winding, as the applied voltage is small and the current to be handled is large. This is because the phase voltage is the same as line to line voltage and the phase current is equal to line current divided by $\sqrt{3}$. The grounding of the neutral of the secondary of a delta-star transformer does not introduce any problem because of third harmonic, as the third harmonic component of the exciting current can flow in the primary delta winding.

When star-delta transformer is used as a step down transformer, the triple harmonic component of exciting current can't flow in the primary winding but it appears in the delta connected secondary winding. In other words, the main winding takes the place of the tertiary winding which is discussed later in connection with star-star windings. It is, therefore, always preferable to have at least one delta connected winding in a three phase transformer, which will eliminate the third harmonic current in the external circuit and thus avoid the interference of power lines with the communication networks.

It is to be noted that if the flux in a transformer magnetic circuit is sinusoidal, the exciting current must contain a third harmonic component. However, because of transformer connections or system connections, this current can't flow, the flux will contain third harmonic component.

2.1.4 Transformer Losses

The various losses in a transformer are enumerated below :
  (i) **Core losses** : These are hysterisis and eddy current losses resulting from alteration of magnetic flux in the core.
  (ii) **Copper loss (I^2R)** : This loss occurs in copper windings when current flow through them.
  (iii) **Load (stray) loss** : It results from leakage fields including eddy currents in the tank wall and conductors.
  (iv) **Dielectric loss** : No dielectric material is perfect, hence, a little leakage current flows through the insulating material. It mainly depends up on frequency of electric field.

2.1.5 Transformer Cooling

It is well known that when the current flows through the conducting medium, heat generates due to I^2R loss. In case of transformers also, during loading conditions a lot of heat is generated and the transformer winding and core temperature increases. High temperatures will damage the winding insulation. Small transformers do not generate significant heat and are self-cooled by air convection currents and radiation of heat. Power transformers rated up to several hundred kVA can be adequately cooled by natural convective air-cooling, sometimes assisted by fans. In larger transformers copper winding and iron core are immersed in transformer oil which helps in both cooling and insulating the windings. The oil is a highly refined mineral oil that remains stable at high temperatures. The oil-filled tank often has radiators through which the oil circulates by natural convection; some large transformers employ forced circulation of the oil by centrifugal pumps. Oil-filled transformers undergo prolonged drying processes to ensure that the transformer is completely free of water vapor before the cooling oil is introduced. This helps prevent electrical breakdown under load. Oil-filled transformers may be equipped with Buchholz relays, which detect gas evolved during internal arcing and rapidly de-energize the transformer to avert catastrophic failure.

Today, non-toxic, stable silicone-based oils, or fluorinated hydrocarbons may be used where the expense of a fire-resistant liquid offsets additional building cost for a transformer vault. Some "dry" transformers (containing no liquid) are enclosed in sealed, pressurized tanks and cooled by nitrogen or sulfur hexafluoride (SF_6) gas.

2.1.6 Transformer Protections

1. **Buchholz Relay :** This relay detects the gas evolution due to internal trouble and gives alarm. It is suited to the detection of minor or slowly developing faults. When a major fault suddenly takes place inside the transformer, oil flows suddenly from the transformer tank to the conservator. The second stage of Buchholz relay is actuated by the oil flow.

2. **Pressure Relief Device :** If the internal pressure of the transformer rises above the certain set value, the pressure relief device starts functioning. It also functions, though occasionally, by the chocking of the air breather. The self excited pressure relief plate is subjected to brittleness caused by secular change. Comprehensive judgment based on the inspection of other protective relay is the key to determine whether the function of this device is an erroneous operation or a normal operation caused by internal defect.

2.2   EXPERIMENTS BASED ON TRANSFORMER

EXPERIMENT NO. 1

2.2.1   To Perform No-Load and Short Circuit Test on a Single Phase Transformer and from the Test Data

(i) Determine the equivalent circuit parameters and draw the equivalent circuit.

(ii) Determine the efficiency and regulation at full-load and 0.8 power factor lagging.

(iii) Draw the efficiency versus load curve.

Necessity of the Experiment

In the electricity generating plant a few kVs like 11, 66 kV or 132 kV can be generated due to insulation problem. But due to economic aspects of transmission of electrical energy higher voltages are used such as 400, 750, 1000, 1500 kVs. For distribution purpose of electrical energy to the consumers, again lower voltages are required for safety point of view. To increase and decrease these voltage levels large transformers are needed.

The performance of large transformers could be determined using direct load test or indirect methods. In direct load test the following problems arise :

(i) A huge energy is wasted

(ii) Large loads are required.

To avoid the above difficulties indirect methods are used for obtaining the performance characteristics.

Theory

This is an indirect method of finding out the performance of large transformers. To obtain equivalent circuit parameters and regulation of transformer two tests are performed- first is no-load and second is short circuit test. The no-load test is generally performed on low voltage (LV) side, because the no-load current I_o is very small (usually 2–10 % of its rated value). Since copper loss ($I^2 R$) is proportional to the square of the current, hence it is negligible and no-load power is approximately equal to the iron-losses (sum of hysterisis and eddy current losses). The equivalent circuit under no-load condition is drawn in Figure 2.7(a) and phasor diagram in Figure 2.7(b).

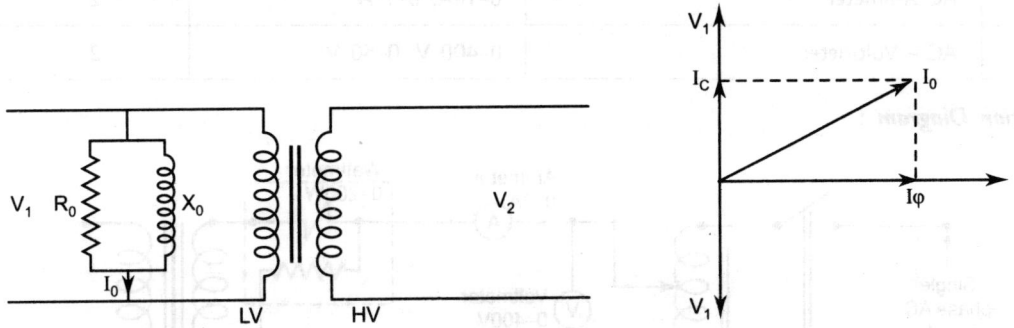

(a) Equivalent circuit of transformer under no-load test   (b) Phasor diagram of transformer under no-load test

FIGURE 2.7

The short circuit test is usually performed on high voltage (HV) side. In this test HV-winding voltage is generally kept low (such as 10 to 20% of rated voltage). Hence, core loss which depends on the voltage is very small and neglected. Thus, power measured in short circuit test is approximately equal to copper-loss. The equivalent circuit and phasor diagram under this test are shown in Figure 2.8.

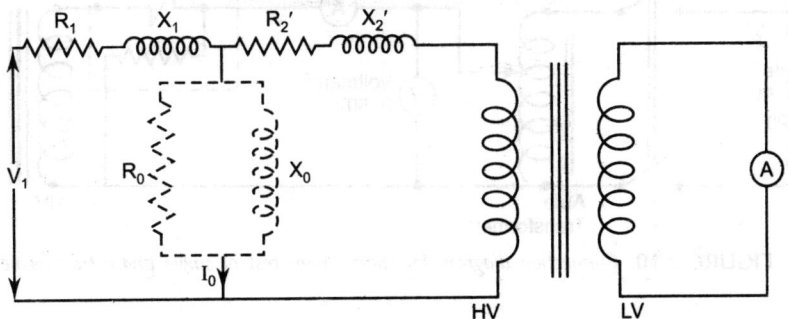

(a) Equivalent circuit of transformer under short circuit test

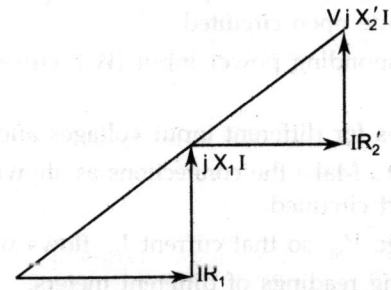

(b) Phasor diagram of transformer under short circuit condition

FIGURE 2.8

Equipments and Components Required

S. No.	Name	Range	No.
1.	Single phase Transformer	400 : 230 V	2
2.	Auto transformer	0–400 V	1
3.	Watt meters	0–200 W, 0–600 W	2
4.	AC Ammeter	0–10A, 0–1 A	2
5.	AC – Voltmeter	0–400 V, 0–50 V	2

Connection Diagram :

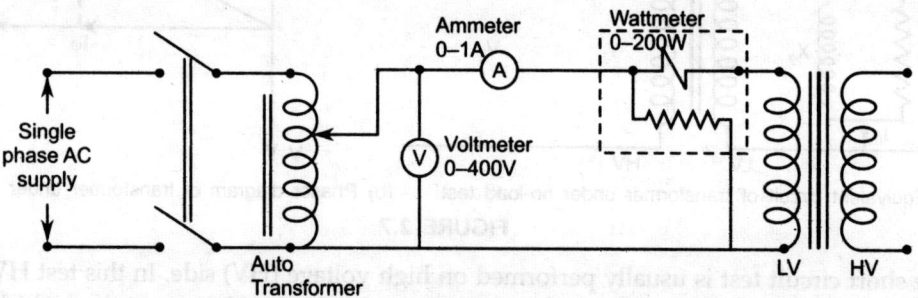

FIGURE 2.9. *Connection diagram for open circuit test of single phase transformer*

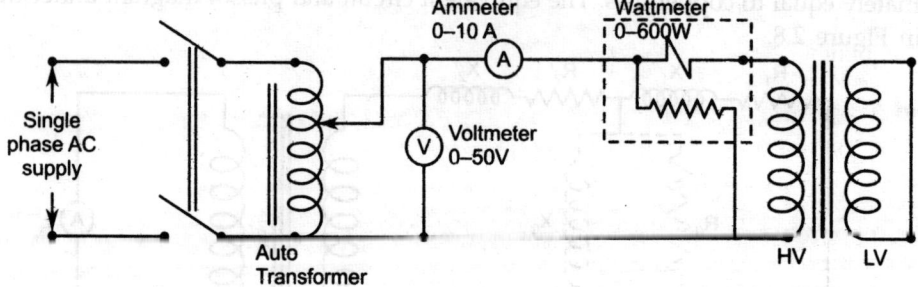

FIGURE 2.10. *Connection diagram for short circuit test of single phase transformer*

Procedure

1. Make the connections as shown in Figure 2.9 for open circuit test. Apply rated voltage on LV side and HV side open circuited.

2. Note down the corresponding power input (W_o), current drawn (I_o) and rated voltage (V_o).

3. Repeat the above steps for different input voltages and tabulate the readings.

4. **For Short Circuit Test :** Make the connections as shown in Figure 2.10 for short circuit test, and LV side short circuited.

5. Apply required voltage V_{SC} so that current I_{SC} flows which is equal to rated current.

6. Note the corresponding readings of different meters.

7. Repeat the above for different short circuit currents and tabulate it.

Observations

For Open Circuit Test

S. No.	Applied voltage V_o (volts)	Current I_o (A)	Power input P_o (W)
1.			
2.			
3.			
4.			
5.			

For Short Circuit Test

S. No.	Voltage applied V_{sc}	Current I_{sc}	SC Power P_{sc}
1.			
2.			
3.			
4.			
5.			

Specifications of Transformer

(a) Single phase transformer

    kVA Rating   — 3

    Primary Voltage — 400V

    Secondary Voltage — 230V

    Frequency   — 50 Hz

Calculations

(a) *From Open Circuit Test* : From this test the R_o and X_o can be determined

The no load power $P_o = V_o \, I_o \, \cos \phi_O$

No load POWER FACTOR $\cos \phi_O = \dfrac{P_O}{V_o \, I_O}$

$\sin \phi_o = $ ....

Refer Figure 2.11.

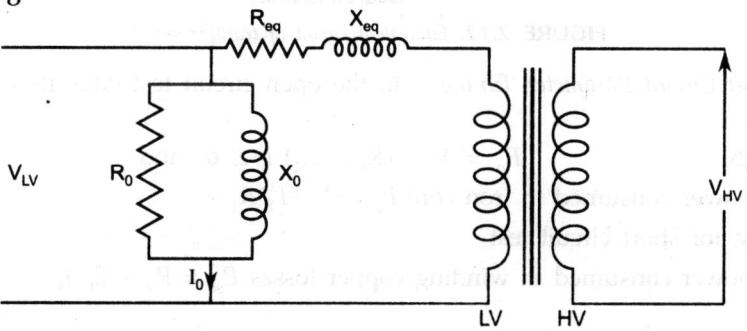

FIGURE 2.11. *Equivalent circuit of transformer*

No-load current I_o has two components core loss component I_c and magnetizing component I_μ.

$$\text{Core loss component } I_c = I_o \cos \phi_0 = \text{................ A}$$
$$\text{Magnetizing component } I_\mu = I_o \sin \phi_0 = \text{................ A}$$

$$R_o = \frac{V_o}{I_c} = \text{................ ohms}$$

$$X_o = \frac{V_o}{I_\mu} = \text{................ ohms}$$

(b) *From Short Circuit Test* : Total impedance referred to HV side

$$Z_{SC} = \frac{V_{SC}}{I_{SC}} = \text{................ ohms}$$

Total resistance referred to HV side

$$R_{eq} = \frac{P_{SC}}{I_{SC}^2} = \text{................ohms}$$

Therefore,
$$X_{eq} = \sqrt{(Z_{sc}^2 - R_{eq}^2)} = \text{................ ohms}$$
$$\text{HV coil reactance } X_1 = X_{eq}/2 = \text{................ ohms}$$
$$n = N_1/N_2 = V_1/V_2$$
$$\text{LV coil reactance } X_2 = X_1/n^2 = \text{................ ohms}$$

Power factor under short circuit test $\cos \phi_{SC} = \dfrac{P_{sc}}{V_{sc} I_{sc}}$

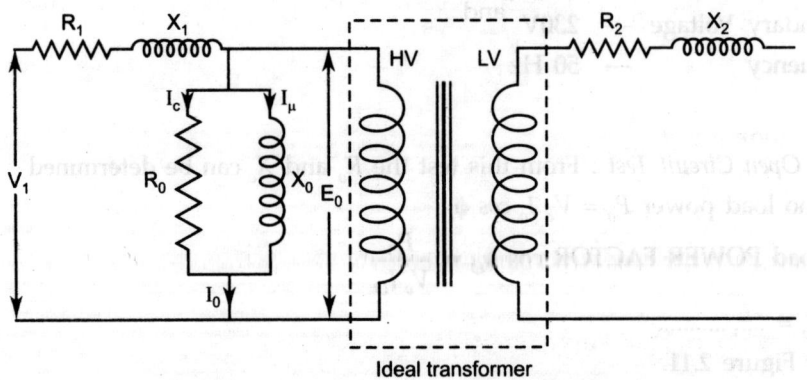

Ideal transformer

FIGURE 2.12. *Equivalent circuit of transformer*

Equivalent Circuit Parameter Tuning : In the open circuit test data, it is considered that $V_1 = E_o$.

Although, $\qquad\qquad E_o = V_1 - (R_1 + jX_1) I_o \angle \phi_O$ and

Actual power consumed in iron core $P_o = P_o - I_1^2 * R_1$.

Similarly, for short circuit test

Actual power consumed in winding copper losses $P_{sc} = P_{sc} - E_0^2/I_{sc}$

(c) *Efficiency Calculation* : Let given power factor be cos ϕ

$$\text{Output} = (kVA) \times 1000 \times \cos \phi = \text{................ Watts}$$

$$\text{Iron loss constant, } W_o = \text{.................. Watts}$$

$$\text{Cu – loss at 'X' time full load} = X^2 \text{ (full load copper loss)}$$

Where $X - \dfrac{\text{Applied load}}{\text{Full load}}$

$$\eta = \frac{\text{Output power}}{\text{Output power + Losses}} \times 100\%$$

(d) *Voltage Regulation* :

$$\frac{V_{NL} - V_{FL}}{V_{NL}} = \frac{\text{Voltage drop}}{V_{NL}} = \frac{I_1 R_1 \cos\phi + I_1 X_1 \sin\phi}{E_1} \quad [V_{NL} \approx E_1]$$

% Voltage regulation $= r \cos\phi + x \sin\phi$

(Positive sign for lagging power factor and negative sign for leading power factor)

Where r – percentage resistance $= \dfrac{I_1 R_1}{E_1} \times 100 = \dfrac{I_1 R_2}{E_2} \times 100$

x – percentage reactance $= \dfrac{I_1 X_1}{E_1} \times 100$

Results

(a) Equivalent parameters

Parameters related with core

$$R_0 = \text{................. and } X_o = \text{.................}$$

HV winding parameters

$$R_1 = \text{................. and } X_1 = \text{.................}$$

LV winding parameters

$$R_2 = \text{................. and } X_2 = \text{.................}$$

(b) Efficiency η = ... %

Always lies between 90 to 100 percent

(c) Regulation = ... %

(d) Draw η verses load curve on graph.

Precautions

1. Perform open circuit test on LV side and HV side is open circuited.
2. Perform short circuit test on HV side and LV side is short circuited.

PROBLEM. *Write a Matlab program to determine transformer equivalent circuit parameters from open-circuit and short-circuit test of single phase transformer. Assign subscript 1 for high voltage winding and subscript 2 for low voltage winding.*

2.2.2 Matlab Program for Equivalent Circuit Parameters

```
%*****************************************************************
%
% tran_eq_ckt.m - Determines transformer equivalent circuit
% parameters.
%
%*****************************************************************
clear;
V1=480; V2=240; n=V1/V2; % Rated voltage values
% Short circuit Test data 'LV side short circuited '
Vsc=37.2; Isc=51.9; Psc=750;
scside='hgh'; %Measurements taken on HV side
% Open circuit Test data 'HV side open circuited'
Voc=240; Ioc=9.7; Poc=720;
ocside='low'; %Measurements taken on LV side
% dc resistance of windings - if not known, set R1dc=R2dc=1
R1dc=0.110; R2dc=0.029;
% Refer all data to HV side
if scside == 'low'
  Vsc=n*Vsc; Isc=Isc/n;
else; end
if ocside == 'low'
   Voc=n*Voc; Ioc=Ioc/n;
else; end

Req=Psc/Isc^2 % Parameters referred to HV side
Zsc=Vsc/Isc
Xeq=sqrt(Zsc^2-Req^2)
R1=Req*R1dc/(R1dc+n^2*R2dc);
X1=Xeq/2

thetaoc=acos(Poc/(Voc*Ioc));
Rc=Voc/(Ioc*cos(thetaoc));
Xm=Voc/(Ioc*sin(thetaoc))
R2=(Req-R1)/n^2; %Parameter referred to LV side
X2=X1/n^2;

disp(' '); disp([' TRANSFORMER EQUIVALENT CIRCUIT PARAMETERS - ',...
   date]);
disp(' '); disp(' ')
disp([blanks(3) 'R1(ohm)' blanks(8) 'R2(ohm)' blanks(8) 'X1(ohm)' ...
   blanks(8) 'X2(ohm)']);
disp([blanks(3) num2str(R1) blanks(9) num2str(R2) ...
   blanks(8) num2str(X1) blanks(9) num2str(X2)]);
disp(' ');
disp([blanks(3) 'Rc(ohm)' blanks(8) 'Xm']);
disp([blanks(4) num2str(Rc) blanks(8) num2str(Xm)]);
% Refinement of data reduction
thetaoc=acos(Poc/(Voc*Ioc));
```

```
Eoc=abs(Voc-(R1+j*X1)*Ioc*exp(-j*thetaoc));
Poc=Poc-Ioc^2*R1; Qoc=Voc*Ioc*sin(thetaoc)-Ioc^2*X1;
Rc=Eoc^2/Poc;
Xm=Eoc^2/Qoc;
thetasc=acos(Psc/(Vsc*Isc)); Esc=abs(Vsc-(R1+j*X1)*Isc*exp(-j*thetasc));
Psc=Psc-Esc^2/Rc; Qsc=Vsc*Isc*sin(thetasc)-Esc^2/Xm;
Req=Psc/Isc^2; R1=Req*R1dc/(R1dc+n^2*R2dc); R2=(Req-R1)/n^2;
X1=Qsc/Isc^2/2; X2=X1/n^2;
disp(' '); disp([' TRANSFORMER EQUIVALENT CIRCUIT PARAMETERS - ',...
    date]);
disp(' Refined Data ');
disp(' '); disp(' ')
disp([blanks(3) 'R1(ohm)' blanks(8) 'R2(ohm)' blanks(8) 'X1(ohm)' ...
    blanks(8) 'X2(ohm)']);
disp([blanks(3) num2str(R1) blanks(9) num2str(R2) ...
    blanks(8) num2str(X1) blanks(9) num2str(X2)]);
disp(' ');
disp([blanks(3) 'Rc(ohm)' blanks(8) 'Xm']);
disp([blanks(4) num2str(Rc) blanks(8) num2str(Xm)]);
%*********************************************************************
```

PROGRAM OUTPUT

TRANSFORMER EQUIVALENT CIRCUIT PARAMETERS—08-Oct-2008

R_1(ohm)	R_2(ohm)	X_1(ohm)	X_2(ohm)
0.13552	0.035729	0.33024	0.082559

R_c(ohm)	X_m
320	104.0715

TRANSFORMER EQUIVALENT CIRCUIT PARAMETERS—08-Oct-2008

Refined Data

R_1(ohm)	R_2(ohm)	X_1(ohm)	X_2(ohm)
0.13532	0.035677	0.32961	0.082403

R_c(ohm)	X_m
319.1152	103.6881

PROBLEM. *Write a Matlab program for calculating the efficiency and regulation of a given transformer whose equivalent circuit parameters are calculated in the earlier program. Also plot the efficiency and regulation variations with load.*

2.2.3 Matlab Program for Efficiency and Regulation

```
%*********************************************************************
%
% trans_perf.m - Calculates voltages, currents, regulation and
% efficiency for load varying from 0% to 150%
%
%*********************************************************************
```

```matlab
clear all;
R1=0.06; R2=0.015; X1=0.18; X2=0.045; Xm=400; Rc=1200;
V1=240; V2=120; KVA=5; n=V1/V2; R2p=n^2*R2; X2p=n^2*X2;
Load=0:10:150; %Percentage load
PF=input('Load Power Factor( 0->1 ) = ');
if PF ~= 1
    PFS2=input('PF Sense( leading,lagging ) = ','s');
else
    PFS2='unityPF'
end
Pout=1000*KVA*PF*Load/100;
if PFS2 == 'lagging'; thet2=-acos(PF);
elseif PFS2 == 'leading'; thet2=acos(PF);
else thet2=0;
end
I2=KVA*1000/V2*Load/100*exp(j*thet2); E1=(R2p+j*X2p)*I2/n+n*V2;
Im=E1/(j*Xm); Ic=E1/Rc; Io=Ic+Im; I1=I2/n+Io;
V1=(R1+j*X1)*I1+E1;
thet1=angle(V1)-angle(I1); PFin=cos(thet1);
if thet1 > 0; PFS1='lagging';
elseif thet1 < 0; PFS1='leading';
else PFS1='unityPF';end
Pin=real(V1.*conj(I1)); eff=Pout./Pin*100; deg=180/pi;
Zm=j*Rc*Xm/(Rc+j*Xm); Znl=R1+j*X1+Zm;
Vnl=abs(V1*Zm/Znl); Reg=(Vnl/n-V2)/V2*100;
deg=180/pi; angV1=angle(V1)*deg; angV2=angle(V2)*deg;
angE1=angle(E1)*deg; angI1=angle(I1)*deg; angI2=angle(I2)*deg;
angIm=angle(Im)*deg; angIc=angle(Ic)*deg;
disp(' '); disp(' TRANSFORMER PERFORMANCE');
disp(' '); disp(' ')
disp([blanks(4) 'V1mag' blanks(4) 'V1ang' blanks(4) 'I1mag' blanks(4)...
    'I1ang' blanks(4) 'I2mag' blanks(4) 'I2ang']);
[abs(V1)' angV1' abs(I1)' angI1' abs(I2)' angI2']
disp(' ');
disp([blanks(4) 'Pin' blanks(4) 'Pout' blanks(4) 'eff' ...
    blanks(4) '%Reg']);
[Pin'/1000 Pout'/1000 eff' Reg']
plot(Load',eff')
xlabel('% Load')
ylabel('Efficiency')
title('Efficiency variation with load')
pause
plot(Load',Reg')
xlabel('% Load')
ylabel('Regulation')
title('Regulation variation with load')
%******************************************************************
```

PROGRAM OUTPUT

Load Power Factor(0 -> 1) = .8

PF Sense (leading, lagging) = lagging

TRANSFORMER PERFORMANCE

V_1mag	V_1ang	I_1mag	I_1ang	I_2mag	I_2ang
ans =					
240.1200	0	0.6325	–71.5651	0	0
240.7706	0.1071	2.6291	–44.7400	4.1667	–36.8699
241.4220	0.2136	4.7026	–41.2604	8.3333	–36.8699
242.0743	0.3196	6.7827	–39.9124	12.5000	–36.8699
242.7273	0.4250	8.8648	–39.1973	16.6667	–36.8699
243.3812	0.5298	10.9478	–38.7543	20.8333	–36.8699
244.0359	0.6341	13.0312	–38.4529	25.0000	–36.8699
244.6914	0.7378	15.1149	–38.2347	29.1667	–36.8699
245.3477	0.8409	17.1988	–38.0693	33.3333	–36.8699
246.0048	0.9435	19.2827	–37.9396	37.5000	–36.8699
246.6627	1.0456	21.3668	–37.8353	41.6667	–36.8699
247.3214	1.1471	23.4509	–37.7495	45.8333	–36.8699
247.9808	1.2481	25.5350	–37.6777	50.0000	–36.8699
248.6410	1.3485	27.6192	–37.6167	54.1667	–36.8699
249.3019	1.4484	29.7034	–37.5643	58.3333	–36.8699
249.9636	1.5478	31.7877	–37.5188	62.5000	–36.8699

P_{in}	P_{out}	eff	%Reg
ans =			
0.0480	0	0	0
0.4488	0.4000	89.1255	0.2709
0.8506	0.8000	94.0480	0.5422
1.2535	1.2000	95.7323	0.8139
1.6574	1.6000	96.5365	1.0858
2.0624	2.0000	96.9765	1.3582
2.4683	2.4000	97.2310	1.6308
2.8754	2.8000	97.3783	1.9038
3.2835	3.2000	97.4581	2.1771
3.6926	3.6000	97.4927	2.4508
4.1027	4.0000	97.4957	2.7248
4.5140	4.4000	97.4755	2.9991
4.9262	4.8000	97.4382	3.2737
5.3395	5.2000	97.3875	3.5486
5.7538	5.6000	97.3266	3.8239
6.1692	6.0000	97.2573	4.0995

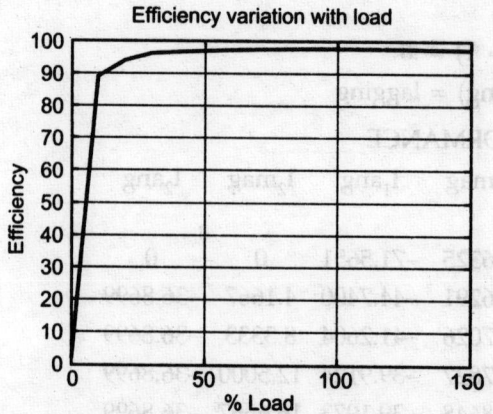

FIGURE 2.13. *Efficiency variation with load*

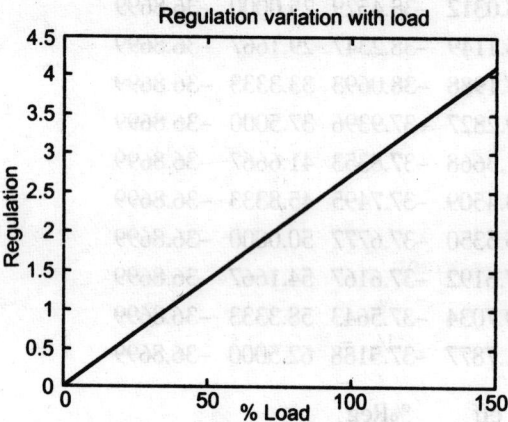

FIGURE 2.14. *Effect of load on voltage regulation.*

EXPERIMENT NO. 2

2.2.4 To Conduct Back to Back Test (Sumpner's Test)

On two identical transformers and from test data :

1. Determine the losses and their equivalent circuit parameters.
2. Determine the efficiency and regulation at full load and 0.8 power factor lagging. Compare the results with the values obtained by short circuit and open circuit test.

Necessity of the Experiment

A load test on a transformer is necessary to find its maximum temperature rise due to different loads. A small rating transformer can be loaded up to full load by means of suitable load impedance. But for large rating transformers full load test is difficult to perform, since it involves considerable waste of energy and a suitable load capable of absorbing full load power is not easily available. However, the performance of large transformers can be determined by means of Sumpner's test.

Equipments and Devices Required

S. No.	Name	Specifications	No. Required
1.	Single phase transformer	400 : 230 V, 50 Hz, 3 KVA	2
2.	Auto transformer	0–400 V	1
3.	AC Voltmeters	0–400 V, 0–230 V	2
4.	AC Ammeter	0–10 A	2

Procedure

1. Make connections as shown in Figure 2.15, this will read either zero or twice the secondary voltage. If it is zero the secondary voltages are in phase opposition.
2. When secondary voltages are in phase opposition, adjust the primary voltage to rated value.
3. Adjust the secondary injected voltage to the required value so that circulating current I_2 is equal to rated secondary current.
4. Note the readings of different meters and tabulate them.

Observations

$$\text{Primary voltage } V_1 = \ldots\ldots\ldots\ldots\ldots$$
$$\text{Primary current } I_1 = \ldots\ldots\ldots\ldots\ldots$$
$$\text{Primary input power } W_1 = \ldots\ldots\ldots\ldots\ldots$$
$$\text{Secondary injected voltage } V_2 = \ldots\ldots\ldots\ldots\ldots$$
$$\text{Secondary current } I_2 = \ldots\ldots\ldots\ldots\ldots$$
$$\text{Secondary power } W_2 = \ldots\ldots\ldots\ldots\ldots$$

Calculations

Iron loss of each transformer $(W_o) = \dfrac{W_1}{2} = \ldots\ldots\ldots\ldots$ watts

No load current of each transformer $I_o = \dfrac{I_1}{2} = \ldots\ldots\ldots\ldots$ A

Full load copper loss of each transformer $W_{SC} = \dfrac{W_2}{2} = \ldots\ldots\ldots\ldots$ watts

Injected voltage per transformer $V_{SC} = \dfrac{V_2}{2} = \ldots\ldots\ldots\ldots$ V

Calculate the equivalent circuit parameters, efficiency and regulation as in short circuit and open circuit tests.

$$R_1 = X_1 = \ldots\ldots\ldots\ldots$$
$$R_0 = X_0 = \ldots\ldots\ldots\ldots$$
$$\text{Efficiency } \eta = \ldots\ldots\ldots\ldots \% \text{ for FL at 0.8 power factor lagging}$$
$$\text{Voltage Regulation} = \ldots\ldots\ldots\ldots \%$$

The results obtained by short circuit and open circuit tests are more accurate than Sumpner's test.

Connecting Diagram

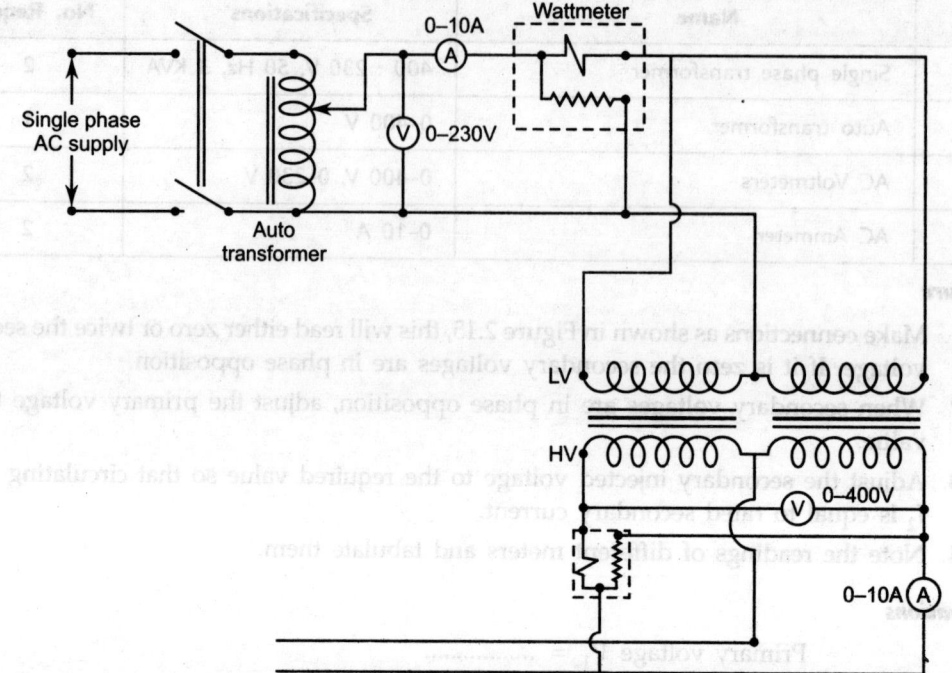

FIGURE 2.15. *Connection diagram for Back to Back or Sumpner's test of single phase transformer*

PROBLEM. *Write a Matlab program to determine the regulation of a transformer for different load power factors.*

2.2.5 Matlab Program to Study the Effect of pf on Regulation

```
%*********************************************************************
%
% trans_reg.m - Regulation of a transformer as a function of power factor
%
%*********************************************************************
clear;
% Equivalent circuit parameters
R1=0.06; R2=0.015; X1=0.18; X2=0.045; Xm=400; Rc=1200;
%Transofrmaer Specifications
V1=240; V2=120; KVA=5;
n=V1/V2; R2p=n^2*R2; X2p=n^2*X2;
npts=100; thet2=linspace(-pi/2,pi/2,npts);
Reg=zeros(1,npts); PF=Reg;
for i=1:npts
    I2=KVA*1000/V2*exp(-j*thet2(i)); E1=(R2p+j*X2p)*I2/n+n*V2;
    Im=E1/(j*Xm); Ic=E1/Rc; Io=Ic+Im; I1=I2/n+Io;
```

```
    V1=(R1+j*X1)*I1+E1;
    Zm=j*Rc*Xm/(Rc+j*Xm);  Znl=R1+j*X1+Zm;
    vnl=abs(V1*Zm/Znl);
    Reg(i)=(Vnl/n-V2)/V2*100;
end
thet2=thet2*180/pi;
plot(thet2,Reg); grid;
title('Regulation vs power factor');
xlabel('Power factor angle ( lead to lag ) - degrees');
ylabel('Regulation, %');
%**************************************************************
```

PROGRAM OUTPUT

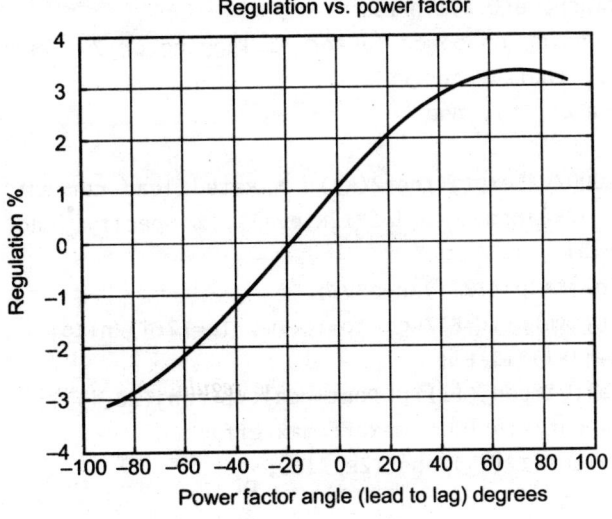

FIGURE 2.16. *Effect of power factor on regulation*

PROBLEM. *Write a Matlab program to study the effect of power factor on the efficiency of transformer at different loads (apparent power). Also determine the maximum efficiency point for different load power factors.*

1. *It is quite clear that the transformer efficiency is zero at no-load and short circuit conditions.*

2. *The efficiency is very low at smaller loads because most of the power input is utilized to overcome core losses.*

3. *As the load increases, efficiency increases and reaches to maximum value, when copper losses become equal to the iron losses. The maximum efficiency point is obtained at nearly 80% of full load. The maximum efficiency is dependent on the cross sectional area of conductor, material used for core and windings, construction of transformers etc.*

4. *Again the efficiency decreases as copper losses become higher than iron losses. Efficiency is also dependent on the power factor of the load applied. At lower power factors efficiency is low for same apparent power and increases as power factor is increased.*

2.2.6 Matlab Program to Study the Effect of pf on Transformer Efficiency

```
%*********************************************************************
%
% Trans_eff.m - Effect of power factor on efficiency of transformer
%
%*********************************************************************
clear all;
% Equivalent circuit parameters
R1=0.09; R2=0.0225; X1=0.18; X2=0.045; Xm=400; Rc=1200;
% Transformer Specifications
V1=240; V2=120; KVA=5;
n=V1/V2; R2p=n^2*R2; X2p=n^2*X2;
nPF=3; PF=linspace(0.2,1,nPF);
PFS2=input('Choice of power factor -1.lagging or 2. leading ? ');
if PFS2 == 1; thet2=-acos(PF);
else; thet2=acos(PF); end
for k=1:nPF
  I2R=KVA*1000/V2*exp(j*thet2(k)); % Rated load current
  nval=150; I2=linspace(0,1.5*I2R,nval); % Specify load values
  for i=1:nval
    E1=(R2p+j*X2p)*I2(i)/n+n*V2;
    Im=E1/(j*Xm); IC=E1/Rc; Io=IC+Im; I1=I2(i)/n+Io;
    V1=(R1+j*X1)*I1+E1;
    Pin=real(V1*conj(I1)); Pout=real(V2*conj(I2(i)));
    eff(i)=Pout/Pin*100; maxeff=max(eff);
    load(i)=abs(I2(i))/abs(I2R)*100;
  end
  for i=1:nval
    if eff(i)==maxeff; nmax=i; else; end
  end
  if k==1; mr='-';
  elseif k==2; mr='-';
  else mr='-.';
  end
  plot(load,eff,mr);
  hold on;
  plot(load(nmax),maxeff,'x')
end
title('Effect of power factor on Efficiency vs load Curve');
xlabel('Load (Apparent Power), %'); ylabel('Efficiency, %');
text(80,25,'Max Eff Point X ');
grid on; hold off;
text(80,35,['- PF = ',num2str(PF(1))]);
text(80,45,['- PF = ',num2str(PF(2))]);
```

```
     text(80,55,['-.- PF = ',num2str(PF(3))]);
        if PFS2==1
    text(80,65,['PF - lagging']);
        else;
        text(80,65,['PF - leading']);
     end
     %*******************************************************************
```

PROGRAM OUTPUT

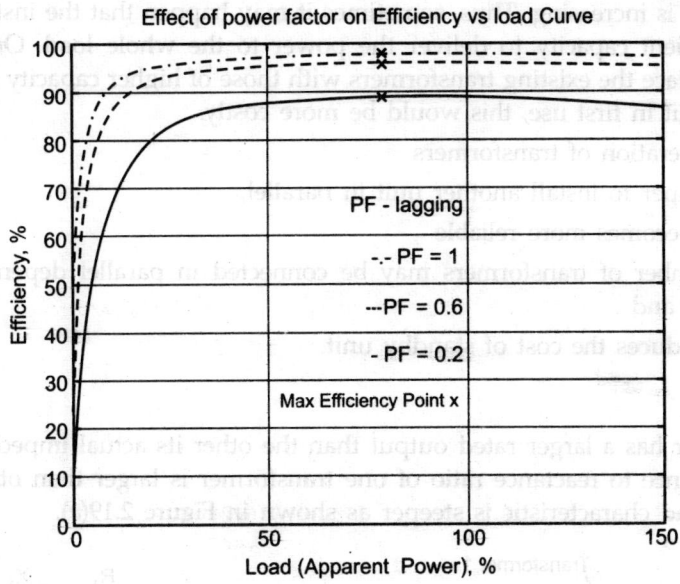

FIGURE 2.17. *Effect of lagging power factor on transformer efficiency*

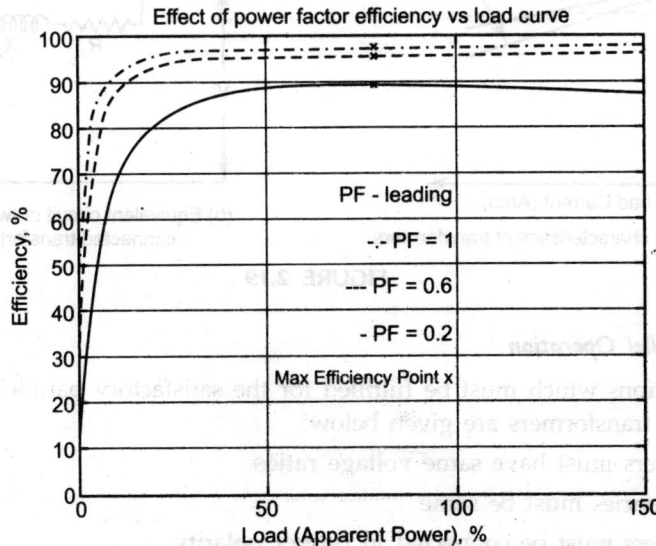

FIGURE 2.18. *Effect of leading power factor on transformer efficiency.*

EXPERIMENT NO. 3

2.2.7 To Perform the Load Test and Parallel Operation of Single Phase Transformer

(i) Draw the load characteristics

(ii) Determine the load sharing by each transformer.

Necessity of the Experiment

The electrical energy is quite clean and easy to control. Due to the increase in population the electrical demand is increasing. Thus, sometimes it may happen that the installed transformer is not having sufficient capacity to deliver the power to the whole load. One way to face this situation is to replace the existing transformers with those of higher capacity or two transformers put in parallel. But in first use, this would be more costly.

In parallel operation of transformers

(i) It is cheaper to install another unit in parallel,

(ii) System becomes more reliable

(iii) Any number of transformers may be connected in parallel depending on the power demand, and

(iv) It also reduces the cost of standby unit.

Theory

If one transformer has a larger rated output than the other its actual impedance will be less. If percentage resistance to reactance ratio of one transformer is larger than other its regulation is poorer and its load characteristic is steeper as shown in Figure 2.19(a).

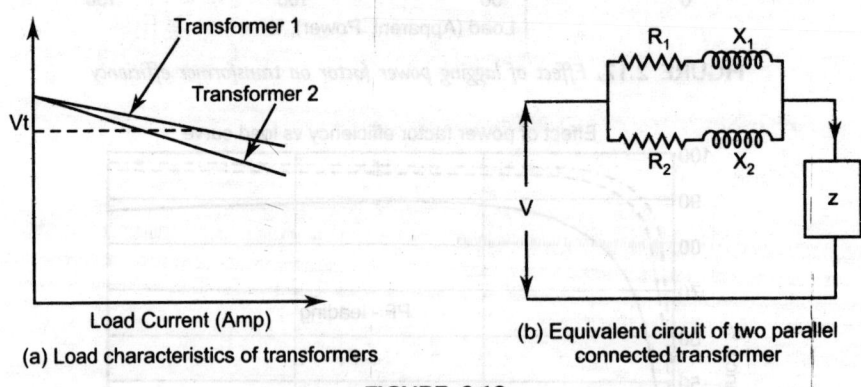

(a) Load characteristics of transformers

(b) Equivalent circuit of two parallel connected transformer

FIGURE 2.19

Conditions for Parallel Operation

The various conditions which must be fulfilled for the satisfactory parallel operation of two or more single phase transformers are given below:

(i) Transformers must have same voltage ratios.

(ii) The frequencies must be same

(iii) Transformers must be connected in proper polarity.

(*iv*) The equivalent leakage impedance in ohms should be inversely proportional to their respective KVA ratings or p.u. impedances must be equal.

(*v*) The ratio of equivalent leakage reactance to equivalent resistance (X_c/R_c) must be same.

Precautions

Condition (*iii*) must be strictly fulfilled. If secondary sides are connected with wrong polarity then large circulating current will flow and transformer may get damaged.

Equipments and Components Required

S. No.	Name	Specifications	No. Required
1.	Single phase transformer	400:230 V, 3 kVA, 50 Hz,	2
2.	Auto transformer	0–300 V	1
3.	AC Voltmeters	0–500 V	1
4.	AC Ammeter	Two of 0–10 A, 0–25A	3
5.	Load trolley		

Connection diagram for load test on single phase transformer

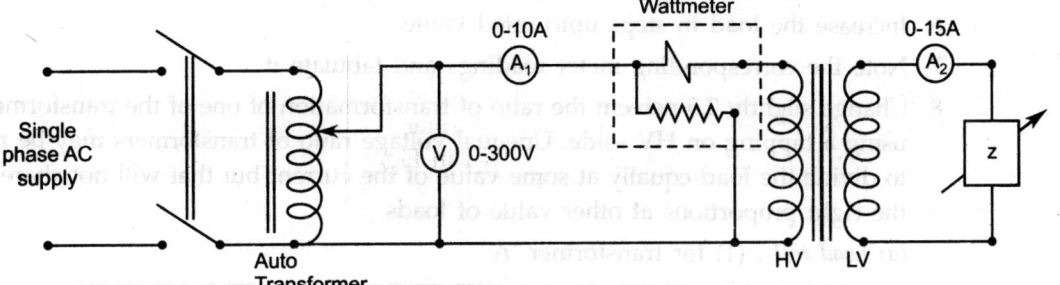

FIGURE 2.20. *Connection diagram for load test of single phase transformer*

Connection diagram for connecting the transformers in proper polarity

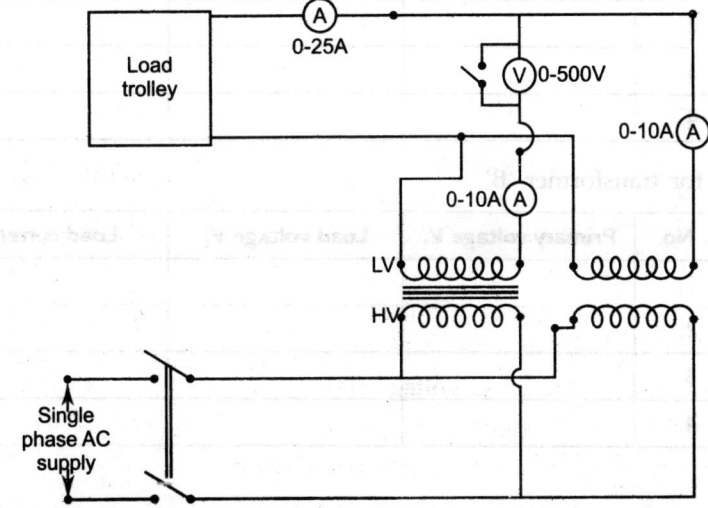

FIGURE 2.21. *Connection diagram for parallel operation of two single phase transformers*

Procedure

(A) For load test :

1. Make the connection as given in Figure 2.20 for conducting the load test.
2. Adjust primary voltage to the rated value.
3. Then load the transformer upto its full load value in steps.
4. Note the reading of load voltage and load current in each step.
5. Repeat the above method for other transformer.

(B) For parallel operation :

1. Make the connection as shown in Figure 2.21.
2. Switch on the supply and note down the voltmeter reading. If voltmeter reading is zero it means two secondary voltages are in phase opposition, then, connect the secondary windings for parallel operation.
3. If reading of voltmeter is sum of the secondary rated voltages i.e., secondary windings are in phase then reverse the secondary terminals of any one transformer.
4. Then connect the meters for taking different readings as shown in Figure 2.21 without disturbing the polarity of either transformer.
5. Adjust the primary input voltage to the rated value.
6. Increase the load in steps upto rated value.
7. Note the corresponding meter readings and tabulate it.
8. Change slightly 2.5 percent the ratio of transformation of one of the transformer by using a tapping on HV –side. Unequal voltage ratio of transformers may be made to divide the load equally at some value of the current but that will not share it in the right proportions at other value of loads.

(a) *Load test* : (1) for transformer 'A'

S. No.	Primary voltage V_1	Load voltage V_2	Load current I_2
1.			
2.			
3.			
4.			

(2) for transformer 'B'

S. No.	Primary voltage V_1	Load voltage V_2	Load current I_2
1.			
2.			
3.			
4.			

(b) *Parallel operation*

S. No.	Load voltage V_1	Load current I	Load shared by 'A' I_1	Load by 'B' I_2	Shared $\frac{I_A}{I_B}$
1.					
2.					
3.					
4.					

Analysis

1. From load test data plot load characteristics for individual transformers.
2. Compare the load shared by each transformer from practical results and results obtained form load characteristics.

Advantages of Parallel Operation

1. The power system becomes reliable *i.e.,* if one transformer fails then it can be removed and other can take load.
2. The transformers can switch on and off depending on power demand. Hence, system becomes more economical.
3. The cost of spare unit is less.
4. It is easy to repair any one transformer without shutting down the power.

EXPERIMENT NO. 4

2.2.8 To Study the Performance of Transformers

On no load and full load for the following connections :
1. Scott connected transformers
2. Open delta or V-connected transformers

Necessity of the Experiment

There are various reasons which are necessary to transform either three–phase to two-phase or vice-versa :

(a) To give a supply to an existing two-phase loads from a new three-phase source.

(b) To supply two phase furnace from a three-phase source.

(c) To supply three-phase motor from a two-phase source.

(d) To supply three- phase or six-phase converter.

(e) To inter line a two-phase system with a 3-phase system for rectifier.

(f) To supply a single-phase load which may or may not be desirable into separate loads and at the same time to maintain reasonable balance on the 3-phase source?

A Scott connection of transformer is used to transform three-phase to two phase and vice-versa.

Theory

(*a*) **Scott Connected Transformers :** In this connection two single phase transformers are connected in such a way that we can supply three- phase power to the set. This connection is used to convert three- phase to two- phase and vice-versa. Two transformers are required for it. One is main transformer having a central tapping and other is teaser transformer having tapping on $0.87 N_1$. Connections are made as shown in Figure 2.22. The secondary voltages of main and teaser transformer are in quadrature and equal in magnitude. Hence these low voltages will constitute a balanced two phase system.

Voltage across primary winding of main transformer = E

Voltage across primary winding of teaser transformer = $0.87\ E$

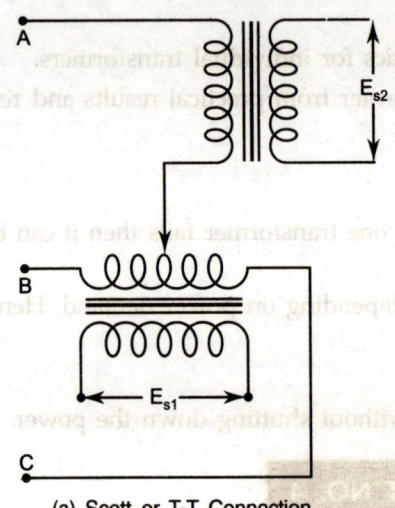

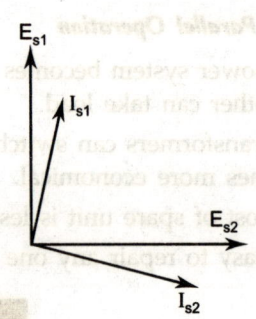

(a) Scott or T-T Connection   (b) Phasor Diagram for balanced inductive load

FIGURE 2.22

Let
$$E_{S1} = \text{Secondary induced voltage for } S_1$$
$$E_{S2} = \text{Secondary induced voltage for } S_2$$
$$I_{S1} = \text{Secondary current in } S_1 \text{ phase at power factor } \cos \phi$$
$$I_T = \text{Teaser transformer primary current}$$

Neglecting magnetizing current in primary windings.

$$I_T\left(\sqrt{\frac{3}{2}} \times T_1\right) = I_2 T_2$$

$$I_T = \sqrt{\frac{2}{3}} I_{S_1} \frac{T_2}{T_1} = 1.18$$

(*b*) **V-Connected Transformer :** The open delta connection or V-connection usually state that except for the voltage unbalancing caused by the unequal impedance of different phases, a balanced three phase load can be supplied with a corresponding balanced line currents and voltages through the loading of the transformer must receive special attention.

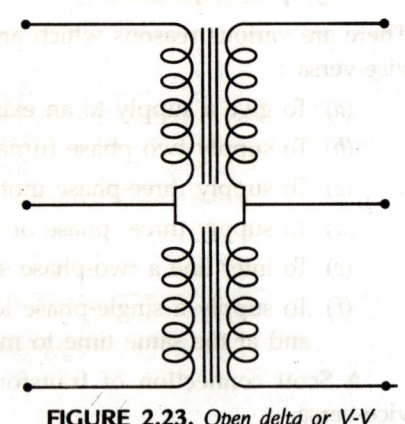

FIGURE 2.23. *Open delta or V-V connection*

When a three-phase alternator is supplying a delta connected transformer bank but due to any cause one transformer gets burned or removed for repair, it becomes an open Δ or V connected transformer bank as shown in Figure 2.23 and continues to supply the power to load.

Let $\qquad$ I – Line current in closed at unity power factor

$\dfrac{I}{\sqrt{3}}$ Phase current

In case of open delta the line current has to be reduced to the phase current. Thus, there is a reduction in power by 42.3%.

Δ-Connected transformer bank capacity $= \sqrt{3}\,EI$

V-Connected transformer bank capacity $= \sqrt{3}\,E = EI$

$$\frac{V-\text{capacity}}{\Delta-\text{capacity}} = \frac{EI}{\sqrt{3}\,EI} = 57.7\ 58\%$$

Disadvantages :

1. For large loads the secondary terminal voltage tends to become unbalanced although your load is perfectly balanced in case of V-connected transformer.
2. Expect unity power factor load two transformers in the V-V bank operates at different power factor.

Connection Diagram

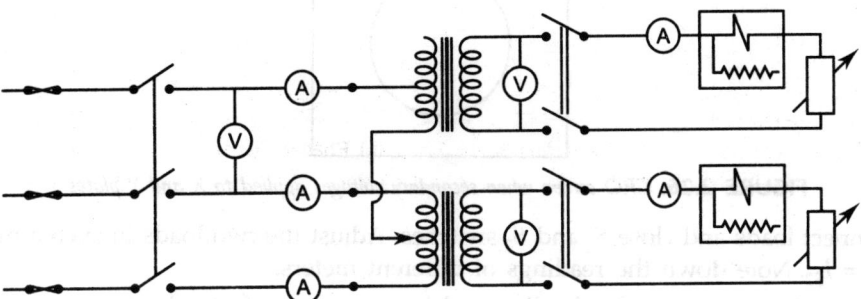

FIGURE 2.24. Load test on Scott connected transformer

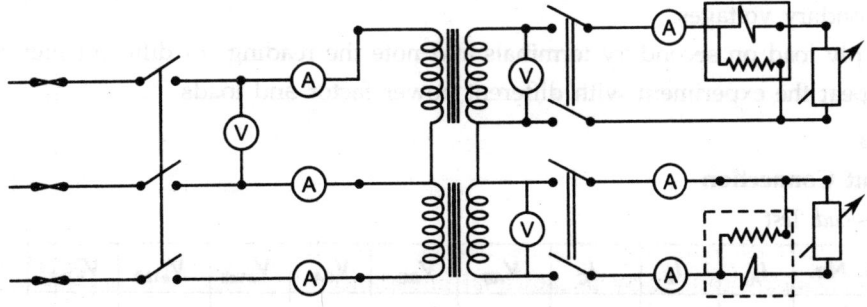

FIGURE 2.25. Load test on V-V connected transformers

Equipments and Components Required

S. No.	Name	Specifications	No. Required
1.	1-phase Transformer	Tapping on 0–220, 381, 440V on primary	2
2.	AC Ammeter		
3.	AC Voltmeter	0–440 V	3
4.	Wattmeter	Single phase	3
5.	Load	Single phase or two phase	1

Procedure

(A) Scott Connection

1. Make the connections as shown in Figure 2.24. Close the switch S_1 and note down the primary and secondary voltages.
2. Feed two secondary voltages to CRO and see the waveforms on CRO. If one voltage is given on X-and other on Y-plates, the resulting Figure 2.26 on CRO will be a circle.

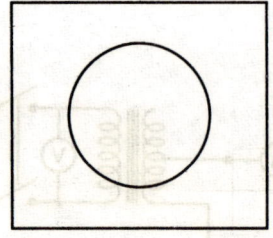

FIGURE 2.26. CRO screen when secondary voltages applied to X and Y plates

3. Connect loads and close S_1 and S_2 switches. Adjust the two loads in such a manner that $I_{S1} = I_{S2}$. Note down the readings of different meters.
4. Repeat the experiment for leading and lagging power factor loads.

(B) V-Connection

1. Make the connections shown in Figure 2.25. Close switch S_1 and note primary and secondary voltages.
2. Apply load on secondary terminals and note the readings of different meters.
3. Repeat the experiment with different power factor and loads.

Observations

(A) Scott Connection

(i) *No-load Test*

S. No.	I_A	I_B	I_C	V_{AB}	V_{BC}	V_{CA}	V_{a1a2}	V_{b1b2}	V_{b2b2}
1.									
2.									
3.									
4.									

(ii) *Load Test*

S. No.	I_{AB}	V_{BC}	V_{CA}	I_B	I_B	I_C	I_{a1a2} I_{b1b2}	V_{a1a2} V_{b1b2}	W_{a1a2} W_{b1b2}	ϕ_2
1.										
2.										
3.										
4.										

(B) *V-V-Connection*

(i) *No-load Test*

S. No.	I_A	I_B	I_C	V_{AB}	V_{ab}	V_{bc}	V_{ca}	I_a	I_b	I_c	$W_1 + W_2$
1.											
2.											
3.											
4.											

(ii) *Load Test*

S. No.	V_{AB}	V_{BC}	V_{CA}	I_A	I_B	I_C	I_a	I_b	I_c	$W_1 + W_2$
1.										
2.										
3.										
4.										

Results

1. In Scott connection when two secondary voltages are given to CRO X and Y plates, a circle is displayed.
2. Similarly in *V*-Connection CRO output is an ellipse.
3. Discussed the currents for balanced and unbalanced load conditions.

EXPERIMENT NO. 5

2.2.9 To Study the Different Standard Connections of Three Phase Transformer

Theory

Three phase circuits are most economical for a.c. power transmission and distribution. As a consequence, three phase transformers are the most widely used in power systems. A three phase transformer may be a single unit (*i.e.*, all windings wound around the same core, immersed in one tank) or may be made up of three phase units (consisting of three separate single phase transformers). In practice the choice between one or another is governed mainly by economical reasons, transportation, future expansion and reliability etc.

The basic type of connections (Y or Δ) of three phase circuits are illustrate in Figure 2.27 and 2.28. The primary and secondary windings of a three phase transformer may be connected in any combination of basic connections.

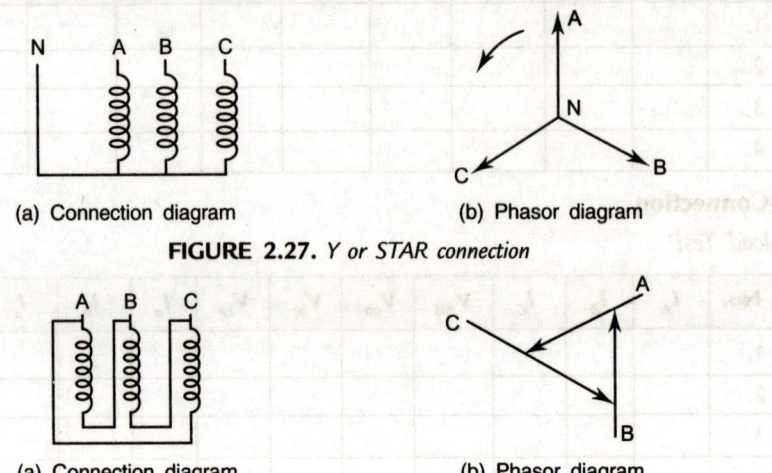

(a) Connection diagram   (b) Phasor diagram

FIGURE 2.27. *Y or STAR connection*

(a) Connection diagram   (b) Phasor diagram

FIGURE 2.28. Δ *or delta connection*

The angular displacement of a transformer connection may be defined as the angle between the no-load line to line voltages of the corresponding primary and secondary terminals. There are two basic angular displacements: 0° and 30° as shown in Figure 2.29. Other angular displacements of 150°, 180°, 210°, 330° (and –30°, –150°), etc., can also be obtained by suitable connections of windings. The importance of angular displacement become clear when three-phase transformers are to be connected in parallel, otherwise short circuit currents will circulate in the secondary windings and damage the transformer.

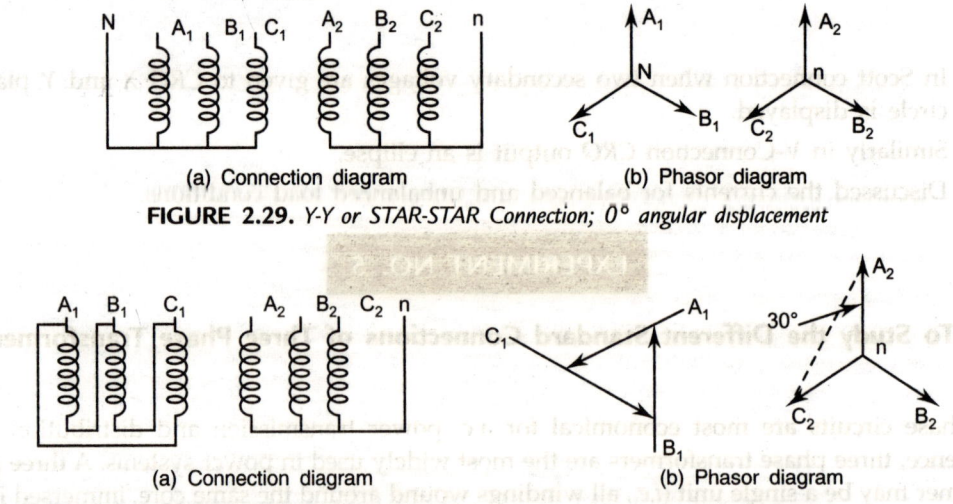

(a) Connection diagram   (b) Phasor diagram

FIGURE 2.29. *Y-Y or STAR-STAR Connection; 0° angular displacement*

(a) Connection diagram   (b) Phasor diagram

FIGURE 2.30. Δ-Y *or delta-star connection; 30° angular displacement*

Due to non-linearity of iron core, higher order harmonics are present in the magnetizing current. Thus, they can be detected in the no-load current. The most significant of them is the third harmonics. It can be shown that third harmonic in all phases is not shifted (is in phase). It can be represented as an independent, superposed current in each of three transformers. Since third harmonic is in-phase and it is needed to generate the sinusoidal flux, some configurations that prevent its flow give non-sinusoidal flux and transformed voltage that in turn contains the third harmonics as sown in Figure 2.31. Such a configuration is for example Y/Y connection with the disconnected neutral.

Let N be the ratio between line to line voltages

In   delta-connection $V_p = V_L$

Star-connection $V_L = \sqrt{3}\, V_p$

$$\text{Star-Star connection } N = \frac{240 * \sqrt{3}}{120 * \sqrt{3}} = 2$$

$$\text{Star-Delta connection } N = \frac{240 * \sqrt{3}}{120} = 2\sqrt{3}$$

$$\text{Delta-delta connection } N = \frac{240}{120} = 2$$

For a three-phase operation, we can interconnect 3 single -phase transformers to form a three phase unit or use a single three-phase unit. With the former arrangement, there is greater flexibility. It is possible to take out one transformer out of service for maintenance or repair and still continue to operate in three-phase mode. The down side is that the *VA* output is reduced.

While inter-connecting three single-phase units to form a three-phase unit, it is necessary to observe polarity marking before inter-connection into Star or Delta.

NOTE ▶ A transformer is an ac device. Most of the transformer impedance comes from the reactive (inductive) component. The resistive component is very small. If you connect it to a dc source, you will fry the windings. Please ensure that **never give dc supply to transformer**.

Common modes of interconnecting single-phase transformers for three phase operation

1. **Y-Y connection :** The second terminal of primary windings of all transformers bunched together to form neutral. Similarly form neutral on the secondary side also. However, it does not permit the flow of third harmonic currents because of the absence of the neutral connection. As a result, the phase voltages will be badly distorted due to third harmonic component. The line to line voltages, however, will be sinusoidal as the third harmonic components will cancel out between the lines. This is seldom used because of problems with 3rd harmonic voltages. The Y connection reduces the amount of insulation needed inside the transformer. The high voltage winding need to be insulated for only $1/\sqrt{3}$ (58%) of the line voltage. There is no phase shift between the incoming and outgoing lines. This connection is shown in Figure 2.29.

2. **Δ-Y connection :** This is used at the sending end to step up the voltage from the generation to the transmission level. The Δ provides a local path for the circulation of third harmonics currents, leading to sinusoidal phase voltages. There is a 30° phase shift between the incoming and outgoing line voltages. This phase difference is undesirable for parallel operation. It results in large circulating currents and reduces the available kVA output from the paralleled transformers. The delta connection provides a path for the circulation of third harmonic currents. Phase voltages will therefore be sinusoidal.

3. **Y-Δ connection :** This is used at the receiving end to step down the voltage from the transmission to the distribution level. Once again there is 30° phase shift between the

incoming and outgoing line voltages. This phase difference is undesirable for parallel operation. It results in large circulating currents and reduces the available kVA output from the paralleled transformers.

4. **Δ–Δ connection :** This is also used frequently. One of the transformers can be taken out of service for repair /maintenance without interrupting continuity of supply. The resulting open delta of *VEE* transformer bank can continue to function in three-phase mode. The downside is that the kVA rating of the reduced to 58% of the original rating.

Harmonics

Very often voltages and currents in power circuits are not perfectly sinusoidal. Usually, the line voltages are very nearly sinusoidal, but the currents may be badly distorted. Distortion may arise from :

1. Magnetic saturation in transformer cores

2. Switching action of thyristors in electronic circuits.

According to Fourier's theorem, any periodic waveform can be decomposed into a fundamental and a series of higher harmonics. The lowest frequency is called the fundamental component. Higher harmonics have frequencies which are integral multiples of the fundamental frequency.
In our Indian system

$$\text{Fundamental frequency} = 50 \text{ Hz}$$
$$2^{nd} \text{ harmonic frequency} = 100 \text{ Hz } (2 \times 50 \text{ Hz})$$
$$3^{rd} \text{ harmonic frequency} = 150 \text{ Hz } (3 \times 50 \text{ Hz})$$

The amplitude of higher harmonics becomes progressively smaller and therefore less important. The predominant harmonics is the 3rd harmonic.

Figure 2.31 shows the flat topped wave produced by the fundamental and third harmonic and its Matlab program is given below.

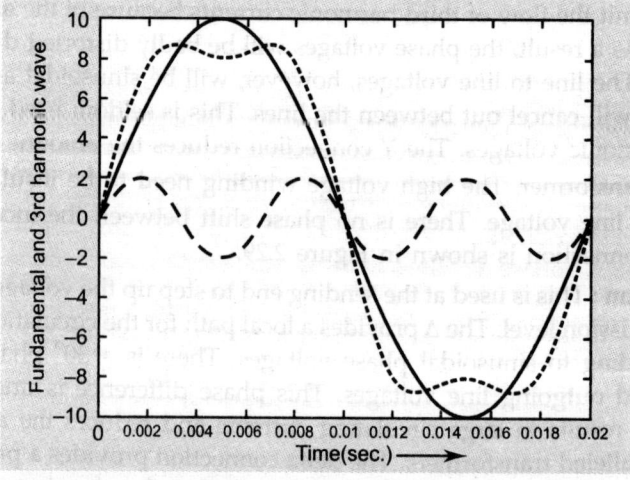

FIGURE 2.31. *Fundamental and third harmonic components*

MATLAB CODE

```
clear all
f=50;                   % Frequency in Hz
w=2*pi*f;               % Frequency in rad/sec
t=0:.0001:.02;          % Time in sec
a1=10*sin(w*t);         % Fundamental wave
a2=2*sin(3*w*t);        % 3rd harmonic component
b=a1+a2;                % Superimposed wave
plot(t,a1,'b-')         % plot
hold
plot(t, a2,'r-')
plot(t,b, 'k:')
xlabel('Time(sec.)-->')
ylabel('Fundametnal and 3rd harmonic wave')
```

Due to magnetic saturation (hysterisis nonlinearity) current and flux (voltage) cannot be sinusoidal at the same time as shown in Figure 2.32. Both could be sinusoidal if *B-H* characteristic is linear (straight line).

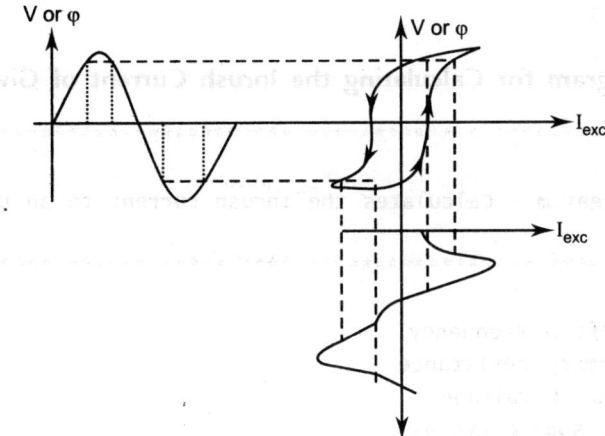

FIGURE 2.32. *Current and flux cannot be sinusoidal simultaneously because of saturation*

If one is sinusoidal, the other would be distorted. Since, we generally desire sinusoidal voltages, for which flux must be sinusoidal. If the current is allowed to be distorted by permitting the flow of 3rd harmonic currents, phase voltages will be sinusoidal. Line to line voltages are always sinusoidal if the phase voltages are distorted. Third harmonic components being in phase, cancel out between lines. **Non-sinusoidal voltages are undesirable since they may cause motors, computers etc., to malfunction.** In star connection of transformers, if neutral point is connected, third harmonic current will be able to circulate. In the delta connection, third harmonic currents can circulate locally in the closed path provided by the delta.

Generally speaking, harmonic voltages and currents are undesirable. But, they are also unavoidable in some ac circuits. Harmonics arise in many ways like :

(*i*) Created by non-linear loads such as saturated magnetic circuits, electric arcs, etc.

(*ii*) Produced by periodic switching of currents and voltages in power electronic circuits.

It is important to understand that only fundamental power does useful work. Harmonic power is dissipated as heat. Therefore, Harmonic currents and voltages are kept as low as possible.

Inrush current of transformer : When a transformer energizes at no-load conditions, the initial transient current flows, which is called inrush current. This current can attain peak levels with values several times the rated value of current. The study of inrush current gets attention for determining the rating of protective breakers. The problem is similar to energizing an *R-L* circuit with sinusoidal source. The steady state current is superimposed on exponentially decaying transient response, necessary to satisfy the initial condition of zero current.

The non-linear differential equation to calculate the inrush current is

$$V_m \sin (\omega t - \xi) \;=\; R_i i_i + \frac{d\lambda}{dt} = R_i i_i + \frac{\partial \lambda}{\partial i}\frac{di}{dt}$$

Where λ – is the flux linkages of the primary winding and $\frac{\partial \lambda}{\partial i}$ is the inductance of the primary winding.

$$\frac{di}{dt} \;=\; -\frac{R_1 i_1}{\partial \lambda / \partial i} + \frac{V_m}{\partial \lambda / \partial i} \sin (\omega t - \zeta)$$

PROBLEM. *Write a Matlab Program for calculating the inrush current of transformer when it is unloaded assuming* $\zeta = 0$.

2.2.10 Matlab Program for Calculating the Inrush Current of Given Transformer

```
%*********************************************************************
%
% inrush_current.m - Calculates the inrush current to an unloaded transformer
%
%*********************************************************************
clear all;
f=50; w=2*pi*f; % Frequency,
R1=0.6; % Primary resistance
Vm=sqrt(2)*230; % Voltage
ncyc=10;% No. source cycles
% No load voltage(rms) and magnetizing current(rms) arrays
Voc=[0 200 240 250 260 270 280 290 300 310 320 480];
Im=[0 0.52 0.65 0.7 0.75 0.85 1 1.5 3 6 12 120]*sqrt(2);
Lam=Voc/(4.448*f); % Flux linkage
np=length(Im);
T=1/f; t0=0; x0=0; x=[x0]; t=[t0];
h=10e-06; incr=T/100; tsav=t0;
for i=1:fix(ncyc*T/h) % Solution for magnetizing current(x)
    if x0 == 0; x0=1e-08; else; end
    if abs(x0) > max(Lam) % Lam-Im slope
        dLamdi=(Lam(np)-Lam(np-1))/(Im(np)-Im(np-1))/sqrt(2);
    else
        a=interp1(Im,Lam,abs(x0)); b=interp1(Im,Lam,1.01*abs(x0));
        dLamdi=(b-a)/(0.01*abs(x0));
    end
```

```
xx=x0; tt=t0;
% Fourth-order Runge-Kutta integration
k1=h*(-R1*xx+Vm*sin(w*tt))/dLamdi;
xx=x0+k1/2;tt=t0+h/2;
k2=h*(-R1*xx+Vm*sin(w*tt))/dLamdi;
xx=x0+k2/2; tt=t0+h/2;
k3=h*(-R1*xx+Vm*sin(w*tt))/dLamdi;
xx=x0+k3; tt=t0+h;
k4=h*(-R1*xx+Vm*sin(w*tt))/dLamdi;
x0=x0+(k1+2*k2+2*k3+k4)/6;
t0=t0+h;
if (t0-tsav) >= incr
tsav=t0; t=[t t0]; x=[x x0];
   else; end
end
plot(t,x); grid;
title('Inrush current');
ylabel('Primary current, A');
xlabel('Time, s');
text(0.7*max(t),0.9*max(x),['Voltage = ',num2str(Vm/sqrt(2))]);
text(0.7*max(t),0.8*max(x),['No. cycles = ',num2str(ncyc)]);
%***********************************************************
```

Results

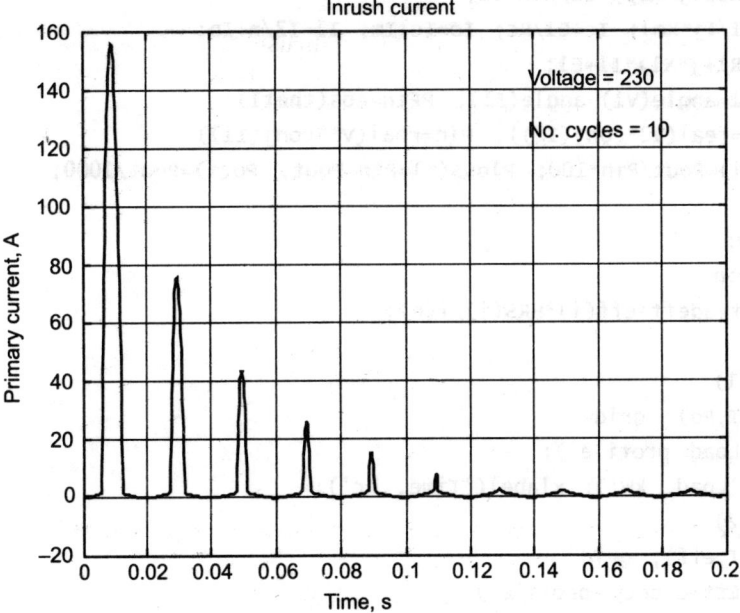

FIGURE 2.33. *Transformer inrush current versus time*

All-day efficiency : In the distribution system the operating cost of transformer is quantitatively assessed with the help of all day efficiency. The all day efficiency depends upon the customer energy use pattern *i.e.,* load profile for 24-hours.

PROBLEM. *Write a Matlab Program for calculating the all day efficiency and average efficiency of a given small size transformer.*

2.2.11 Matlab Program for Calculating the All Day Efficiency of Transformer

```matlab
%***********************************************************************
%
% Allday_eff.m - calculates the average or all day efficiency.
%
%***********************************************************************
clear all;
%Transformer parameters
R1=0.20; R2=0.002; X1=0.60; X2=0.006; Xm=130; Rc=4000;
% Transofrmer specifications
V1=2400; V2=240; KVA=25; n=V1/V2; R2p=n^2*R2; X2p=n^2*X2;
LOAD=[55 75 100 90 65 55]; % Percent apparent power load
PF=[0.8 0.85 0.9 0.85 0.77 0.8]; % Power factor, lagging assumed
HRS=[6 6 4.5 3.5 2 2]; % Hours at load point
n=length(HRS); T(1)=HRS(1); for i=2:n; T(i)=T(i-1)+HRS(i); end
if T ~= 24; disp('WARNING - 24 hour period not specified'); end
for i=1:n
    I2=KVA*1000/V2*LOAD(i)/100/PF(i)*exp(-j*acos(PF(i)));
    E1=(R2p+j*X2p)*I2/n+n*V2;
    Im=E1/(j*Xm); Ic=E1/Rc; Io=Ic+Im; I1=I2/n+Io;
    V1=(R1+j*X1)*I1+E1;
    thet1=angle(V1)-angle(I1); PFin=cos(thet1);
    Pout=real(V2*conj(I2)); Pin=real(V1*conj(I1));
    eff(i)=Pout/Pin*100; Ploss(i)=Pin-Pout; Po(i)=Pout/1000;
end
adeff=0;
for i=1:n
    adeff=adeff+eff(i)*HRS(i)/T(n);
end
figure(1)
stairs(T,Po); grid
title('Load profile');
ylabel('Load, kW'); xlabel('Time, hr');
figure(2)
stairs(T,eff); grid
title('Effieiency profile');
ylabel('Efficiency, %'); xlabel('Time, hr');
disp(' ALL DAY EFFICIENCY STUDY'); disp(' ')
disp([' All-Day Efficiency(%) - ', num2str(adeff)]);
%***********************************************************************
```

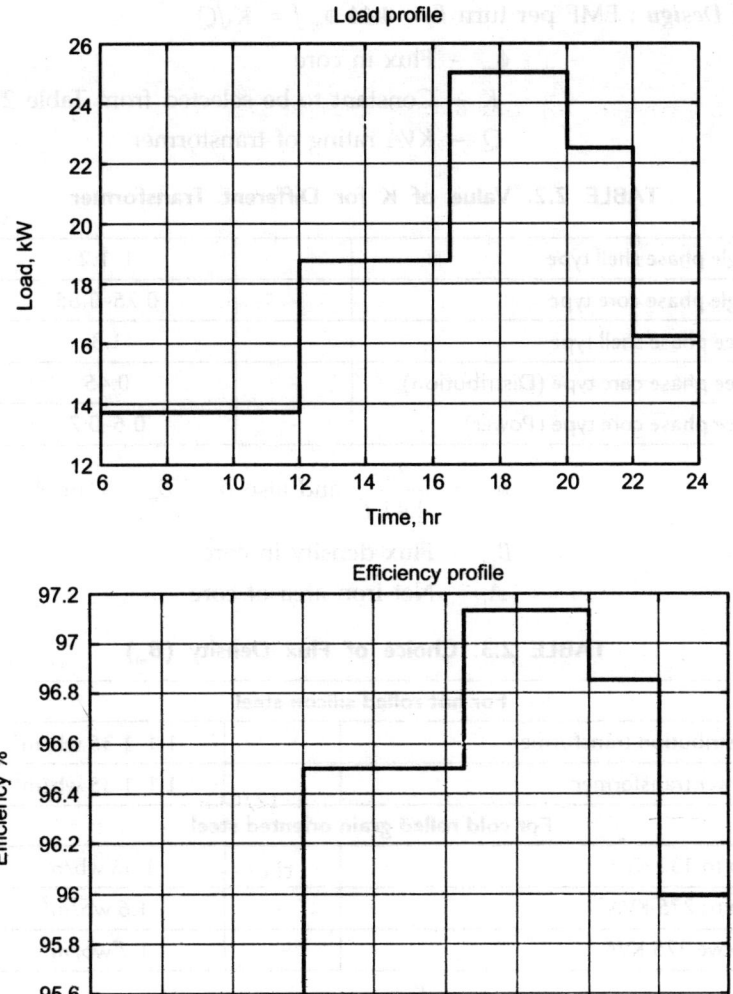

FIGURE 2.34. *Load variation and transformer all day efficiency*

2.2.12 Transformer Design Procedure

Step 1. *Specifications* :

kVA rating	
Voltage rating	
Frequency	
No. of phases	
Connection	
Type of cooling	

Step 2. *Core Design* : EMF per turn $E_t = 4.44 \, \phi_m \, f = K\sqrt{Q}$

ϕ_m – Flux in core

K – Constant to be selected from Table 2.2

Q – *KVA* rating of transformer

TABLE 2.2. Value of K for Different Transformer

Single phase shell type	–	1–1.2
Single phase core type	–	0.75–0.85
Three phase shell type	–	1.3
Three phase core type (Distribution)	–	0.45
Three phase core type (Power)	–	0.6–0.7

$$\phi_m = \frac{E_t}{4.44 \, f} \quad \text{and also} \quad \phi_m = B_m \cdot A_i \quad \text{or} \quad A_i = \frac{\phi_m}{B}$$

B_m – Flux density in core

A_i – Net iron area of core

TABLE 2.3. Choice of Flux Density (B_m)

For hot rolled silicon steel		
Distribution transformer	–	1.1–1.35 wb/m^2
Power transformer	–	1.1–1.35 wb/m^2
For cold rolled grain oriented steel		
Up to 132 KVA	–	1.55 wb/m^1
Up to 275 KVA	–	1.6 wb/m^2
Above 275 KVA	–	1.7 wb/m^2

Net iron area of core $A_i = \dfrac{K_c}{d^2}$

K_c – may be selected from Table 2.4

Step one	Step two	Step three	Step four
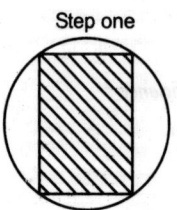	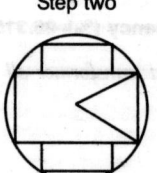	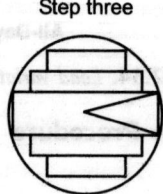	

FIGURE 2.35

TABLE 2.4. Value of K_c for Different Core of Transformer

$A_i = K_c \, d_2$	1-step square	2-step cruciform	3-step	4-step
K_c	0.45	0.56	0.6	0.62

Diameter of circumscribing circle $d = \sqrt{(A_i / K_c)}$

To be determine the step length and width of core cross section

$\theta = 90/n$ n-number of step in core

$\theta = 90/1 = 90$ for one step core

$\theta = 90/2 = 45$ for two step core

$\theta = 90/3 = 30$ for three step core

$\theta = 90/4 = 22.5$ for four step core

On a graph paper draw a diagram showing complete details of core cross section with dimensions and steps.

Step 3. *Window Dimensions* : Select the window space factor (Kw) according to kVA rating of transformer.

$$kW = \frac{8}{(30 + kV)} \quad \text{up to 20 KVA}$$

$$kW = \frac{10}{(30 + kV)} \quad \text{up to 250 kVA}$$

$$kW = \frac{12}{(30 + kV)} \quad \text{up to 1000 kVA and above.}$$

Select the current density (δ) in the winding according to method of cooling :

TABLE 2.5. Value of Current Densities for Different Types of Cooling in Transformers

Natural cooling	AN—Air Natural	$\delta = 1.1$–$1.28 * 10^6$
	ON—Oil Natural	
	OFN—Oil Forced Natural	
Forced cooling	AB—Air Blast	$\delta = 3$–$4 * 10^6$
	OB—Oil Blast	
	OFB—Oil forced Natural	
Forced cooling with water circulating	OW—Oil cooled with water circulation	$\delta = 4.1$–$6 * 10^6$
	OFW—Forced Oil cooled with water circulation	

Output of transformer

$$Q = 3.33 f B_m \delta K_w A_w A_i^* 10^{-3} \text{ kVA} \quad \text{for three phase core type}$$
$$Q = 6.66 f B_m \delta K_w A_w A_i^* 10^{-3} \text{ kVA} \quad \text{for three phase Shell type}$$

Calculate window area (A_w) from above equation

Select the ratio of height (H_w) of the window to the width (W_w) of the window.

$$H_w/W_w = 2 \text{ to } 4$$

and $$A_w = H_w W_w$$

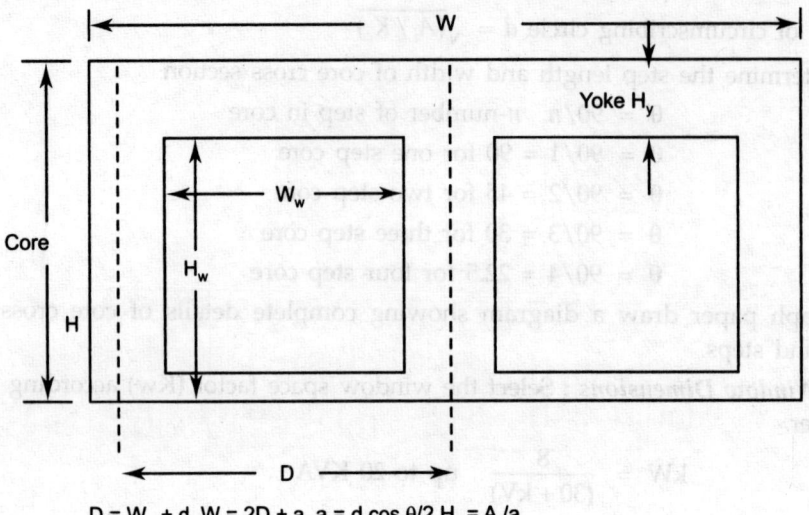

$D = W_w + d$, $W = 2D + a$, $a = d \cos \theta/2$ $H_y = A/a$

FIGURE 2.36

Step 4. *Yoke Design* : The area of the Yoke is taken as 1.2 times that of limb to reduce iron losses.

$$\text{Flux density in yoke } B_y = B_m / 1.2$$
$$\text{Net yoke area } A_y = 1.2 \, A_i$$
$$\text{Depth of the Yoke } = a = d/2 \cos \theta/2$$
$$\text{Height of the yoke } H_y = A_y / a$$

Step 5. *Overall Design of Frame*

$$\text{Height of frame } H = H_w + 2 \, H_y$$
$$\text{Width of frame } W = a + 2 \, D$$
$$\text{Depth of frame } D_y = a$$

Step 6. *LV Winding Design* : LV winding is kept near the core to reduce insulation cost.

$$\text{LV winding Voltage } = V_L$$

Calculate phase voltage for LV winding

$$V_{phL} = V_L / \sqrt{3} \text{ for Y connection}$$
$$V_{phL} = V_L \text{ for } \Delta \text{ connection}$$

$$\text{Number of turns per phase } T_L = V_{phL} / E_t$$

$$\text{LV winding phase current } I_{phL} = kVA \times 1000 / (\sqrt{3} \, V_{phL})$$

Cross sectional area of LV winding conductor $a_L = I_L / \delta$

Select the corresponding width and thickness of copper conductor and make a round figure of A_L up to one place of decimal.

$$\text{Dimension of bare conductors } A_L = x.y \text{ mm}^2$$

Add 0.25 for paper coating on conductor then dimension of insulated conductor

$$= (x + 0.5) \, (y + 0.5) \text{ mm}^2$$

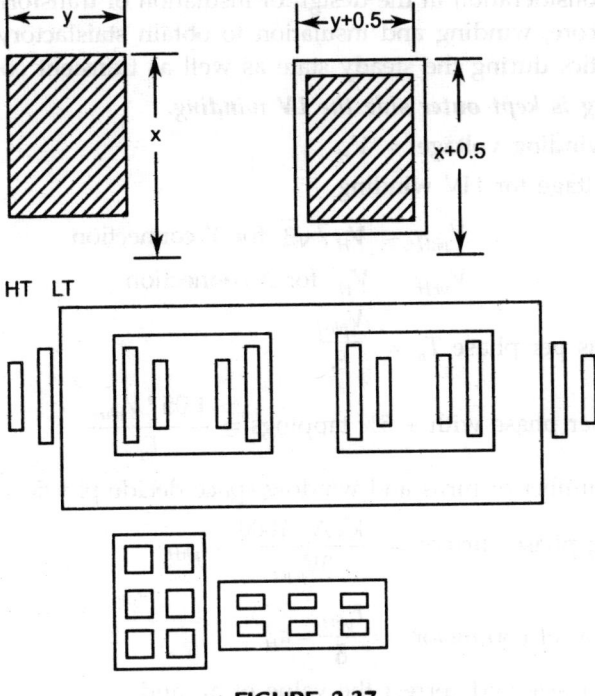

FIGURE 2.37

Number of layers (1, 2, 3....) should be decided by total number of turns in the winding, height and width of window.

   Use   0.5 mm pressboard cylinders between layers

         0.5 mm pressboard

         0.25 mm paper insulation

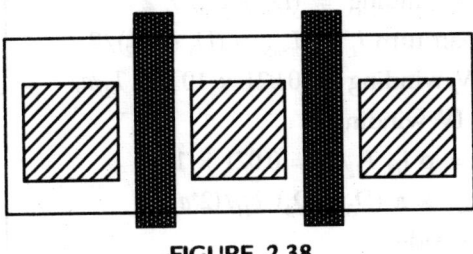

FIGURE 2.38

Axial depth of LV winding - L_{cs}

Radial depth of LV winding = (Number of layers x radial depth of conductor) + insulation between layers

Pressboard is also used between core and LV winding.

Thickness of insulation between core and LV winding and HV and LV winding can be calculated by

$$\text{Thickness} = (5 + 0.9 \text{ kV}) \text{ mm}$$

Thickness of insulation includes width of any oil duct provided in between. Width of oil duct varies from 6 mm – 12 mm depending upon capacity and voltage of transformer.

The fundamental consideration in the design of insulation of transformers may be described as those of arranging core, winding and insulation to obtain staisfactory electrical, mechanical and thermal charateristics during the steady state as well as transient coditions.

Step 7. *HV winding is kept outer side the LV winding.*

HV winding voltage $= V_H$

Calculate phase voltage for HV winding

$$V_{phH} = V_H / \sqrt{3} \text{ for } Y \text{ connection}$$
$$V_{phH} = V_H \text{ for } \Delta \text{ connection}$$

Number of turns per phase $T_h = \dfrac{V_{phH}}{E_t}$

Number of turns per phase with $\pm$ 5% tappings $= \dfrac{1.05 * V_{phH}}{E_t}$

Depending upon number of turns and window space decide physical shape of HV winding.

HV winding phase current $= \dfrac{kVA * 1000}{3V_{phH}} = I_{phH}$

Area of cross section of conductor $= \dfrac{I_{phH}}{\delta} = a_H$

Select the conductor size and correct the value of a_H and

$$\delta_{Modified} = I_{ph}/a_H$$

Like the LV winding determine the axial and radial depth of HV winding.

Insulation between HV and LV winding = thickness of bakellised paper + oil duct.

Determine inside and outside diameter of HV winding.

Step 8. *Resistance*

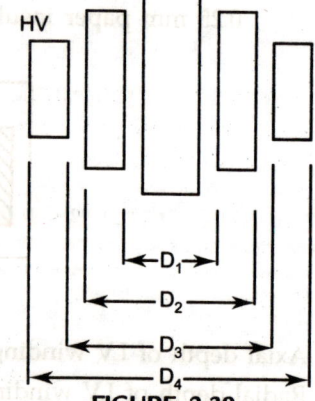

Mean diameter of LV winding $= (D_1 + D_2) / 2$

Length of mean turn $l_L = L_{mtL} = (D_1 + D_2)/2$

Resistance of LV winding $= 0.021 \times 10^{-6} \, l_L \, T_L/a_L$

Similarly resistance of HV winding

$$r_H = \rho \, l_H \, T_H/a_H - 0.021 \times 10^{-6}$$
$$\times \pi \, (D_3 + D_4) \, T_H/(2*a_H)$$

Total resistance ref to LV side

$$R_L = r_L + r_H/K^2, \ K = V_{phH}/V_{phL}$$

P.U. resistance ref to LV side

$$R_{PUL} = R_L \times I_{phL}/V_{phL}$$

FIGURE 2.39

Step 9. *Leakage Reactance* : Leakage reactance of transformer referred to LV side (three phase core)

$$X_L = 2 \pi f \mu_0 \, T_L^2 \, L_{mt} \, (a + (b_p + b_s)/3)/L_c$$

L_c – Average height of windings

$L_{mt} = \pi \, (D_1 + D_2)/2$

A – distance between HV and LV winding

B_p – width of LV winding

B_s – width of HV winding

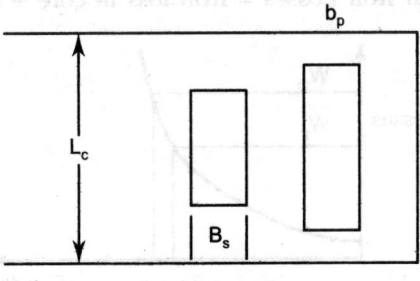

FIGURE 2.40

P.U. leakage reactance $X_{Lpu} = X_L \times I_{phL}/V_{phL}$

$$\text{P.U. Impedance} = \sqrt{\left[(\text{P.U. Resistance})^2 + (\text{P.U. Reactance})^2\right]}$$

Step 10. *P.U. Regulation*

$$\text{P.U. Regulation} = \sqrt{\left[(\text{P.U. Resistance})\cos\phi + (\text{P.U. Reactance})\sin\phi\right]}$$

Step 11. *Losses* : *Copper Losses*

$$\text{Total copper losses} = 3\, I_{phL}^2\, R_L$$

Assume stray losses $= W_{st} = (10 \text{ to } 15\ \%)\ W_{copper\ loses}$

Iron losses :

Density of CRGO stell $= 7.6 \times 10^3\ \text{Kg}/\text{m}^3$

Calculate weight of core and Yoke

Flux density in core $= B_m$

Flux density in Yoke $= B_y$

From Figure 2.41 below, select specific iron losses (watt/Kg) corresponding B_m and B_y.

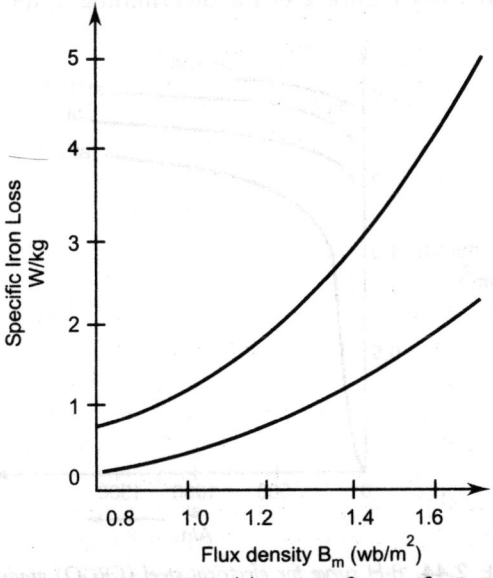

FIGURE 2.41. *Typical loss curve for transformer*

Iron losses in core = W_m × weight of core

Iron losses in yoke = W_Y × weight of yoke

W_i = Total Iron Losses = Iron loss in core + Iron loss in yoke

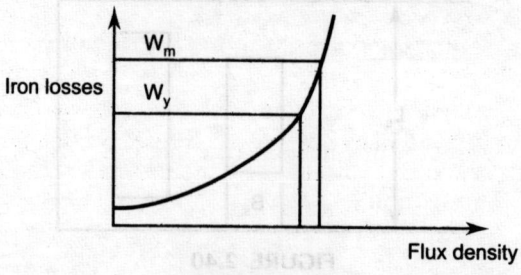

FIGURE 2.42

Step 12. *Efficiency*

$$\text{Total Losses } W_T = W_{copper\ loss} + W_i + W_{st}$$

$$\text{Efficiency } \eta = \frac{kVA \times pf}{(kVA \times pf + W_T)}$$

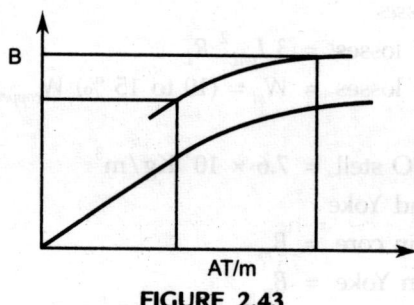

FIGURE 2.43

Step 13. *No load current* : Ref Figure 2.44 for determining amper turn (at).

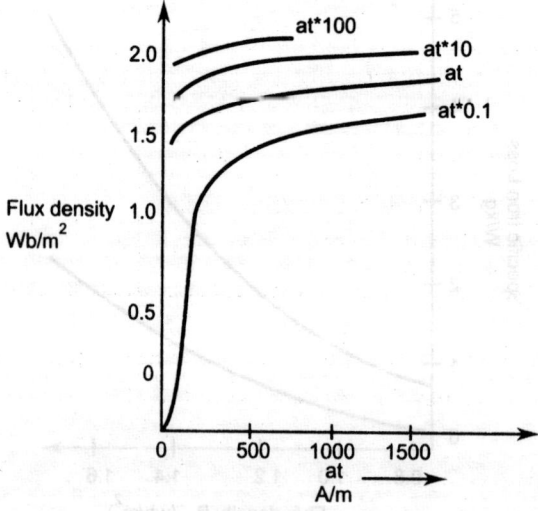

FIGURE 2.44. *B-H curve for electrical steel (CRGO) grade 123*

For B_m and B_y select AT_c and AT_y

Total magnetizing MMF = (total length of core). at_c + (total length of yoke). at_y

Magnetizing MMF per phase AT_0 = (Total magnetizing MMF)/3

Magnetizing current $I_m = AT_0 / (2\ T_L)$

Loss component of no-load current I_1 = Total iron loss/(3* V_{phL})

$$\text{No load current } I_o = \sqrt{(I_m^2 + I_l^2)}$$

I_o should not be greater than 5 % of full load current.

Step 14. *Design of Tank*

Height of Tank H_t = H + base clearance + extra oil level + height of leads

Width of tank W_t = W + diameter of outer winding + clearance for tank wall on both sides

Length of tank = Diameter of outer winding D_4 + clearance for tapping wires on each side

Tubes are provided for cooling of transformer permissible temperature rise of oil = θ

Total area of 4 walls of tank = S_T

Let area of tubes = $X . S_T$

X = (Total loss/(S_T . θ) – 12.5)/8.8

Total area of the tubes A_T = (Total loss/(θ) – 12.5 S_T)/8.8

Number of tubes N_l = $A_T / \pi\ d_t\ l_t$

d_t – diameter of tube = 50 mm

l_t – length of the tube = height of tank

spacing between tubes = 75mm

2.2.13 Matlab Program for Transformer Design

```
% STEP1,2,3,4
% Inputs
clear all
kva=input(' Enter kVA Rating of Transformer ');
lv=input(' Enter Voltage Rating of LV ');
hv=input(' Enter Voltage Rating of HV ');
disp(' If you wish that ');
disp(' Remaining inputs as default ')
disp(' Press - 1 otherwise - Press 0 ');
default= input(' your Choice Please = ');
if default==0
    f=input(' Enter value of frequency \n ');
    cooling_med=input('\n Cooling Medium \n 1. Air \n 2. Oil \n Your Choice = ');
    circulation_type =input('\n Circulation Type of cooling Medium \n 1. Natural
        cooling \n 2. Forced cooling \n Your Choice = ');
    cooling=input('\n Type of cooling \n 1. Simple cooling \n 2. Mixed cooling \n
        Your Choice = ');
    No_Phase=input('\n Number of Phase(s) \n 1. Single phase \n 2. Three phase \n
        Your Choice = ');
```

```
        core_type=input('\n Type of Core \n 1. Core Type \n 2. Shell Type \n
            Your Choice = ');
        phase=1; trans_type=1; connection=1;
        if No_Phase==2
        phase=3;
        trans_type = input('\n Type of Transformer \n 1. Power Transformer \n 2.
            Distribution of Transformer \n Your Choice = ');
        connection=input('\n Type of connection \n 1. star/delta \n 2. delta/star \n
            3. star/star \n 4. delta/delta \n Your Choice = ');
        end
        core_step=input('Select type of core \n 1. Single step core (Square core) \n
            2. Two Stepped cruciform core \n 3. Three stepped core \n 4. Four stepped
            core \n');
        else
        f=50;
        phase=1;
        core_type=1;
        cooling_med=1;
        circulation_type=1;
        cooling=1;
        core_step=1;
        trans_type=1;
        connection=0;
    end
    if (phase==1)& (core_type==1)
        k=.8;
    elseif (phase==1)& (core_type==2)
        k=1.1;
    elseif (phase==3)& (core_type==1) &(trans_type==1)
        k=0.65;
    elseif (phase==3)& (core_type==1)&(trans_type==2)
        k=.45;
    else
        k=1.3;
    end
    e=k*sqrt(kva);
    fi=e/(4.44*f);
    if(trans_type==1)
        bm=1.4;
    else
        bm=1.2;
    end
    ai=fi/bm;
    if(core_step==1)
        kc=.45;
        thita=3.14/2;
```

```
elseif(core_step==2)
   kc=.56;
   thita=3.14/4;
elseif(core_step==3)
   kc=.6;
   thita=3.14/6;
else
   kc=.62;
   thita=(22.5*3.14)/180;
end
   d=sqrt(ai/kc);

if(kva<=20)
   kw=8/(30+hv);
elseif(kva>20 & kva<=250)
   kw=10/(30+hv);
else
   kw=12/(30+hv);
end
if(cooling_med==1) & (circulation_type==1)& (cooling ==1)
   delta=1.15e+6;
elseif(cooling_med==2) & (circulation_type==1)& (cooling ==1)
   delta=1.3e+6;
elseif(cooling_med==2) & (circulation_type==1)& (cooling ==2)
   delta=3e+6;
elseif(cooling_med==2) & (circulation_type==2)& (cooling ==2)
   delta=5e+6;
else
   delta=1.6*1000000;
end
   if (phase==1)& (core_type==2)
      aw=(kva*1000)/(6.66*f*bm*delta*kw*ai);
   else
      aw=kva*1000/(3.33*f*bm*delta*kw*ai);
end
ww=sqrt(aw/3); % width of window
hw=3*ww; % height of window
D=ww+d;
a=d*cos(thita/2);
W=2*D+a;
by=bm/1.2;
ay=1.2*ai;
hy=ay/a;
H=hw+2*hy;
dy=a;

% step5&6
if ((connection==1|connection==3) & phase==3)
   vph1=lv/sqrt(3);
```

```
    else
       vph1=lv;
    end
    if (connection==2|connection==3)
       vphh=hv/sqrt(3);
    else
       vphh=hv;
    end
    t  =round(vph1/e);                   % no. of turns in LV winding
    iph1=kva*(1000/(3*vph1));            % Current in LV winding =kVA/lv
    al=iph1/delta;                       % Area of Copper for LV
    x=sqrt(al/3);                        % Ratio of width to thickness is 1:3
    y=3*x;
    ins_thick_lv=(5+0.9*vph1/1000);      % insulation thickness for LV
    ins_thick_hv=(5+0.9*vphh/1000);      % insulation thickness for HV
    X=x+0.05;
    Y=y+0.05;                            % insulation thickness 0.05 on conductor
    AL=X*Y;                              % Total Area of LV winding
    lcs=tl*Y;                            % Axial Depth of LV winding
    no_layers=3;
    bs=no_layers*X+2*0.5;                % Radial Depth of LV winding
    Dil=d+(2*ins_thick_lv);              % inside dia. of LV winding
    Dol=Dil+(2*9*bs);                    % outside dia. of LV wdg.
    DL=(Dil+Dol)/2;                      % Mean Dia. of LV winding

    th=round(1.05*vphh/e);               % no of turns in HV winding with 5% tapping
    iphh=kva*1000/(3*vphh);              % Current in HV winding
    ah=iphh/delta;                       % Area of HV conductor
    dh=sqrt(4*ah/pi);                    % Diameter of bare conductor of HV
    dh_ins=dh+0.15;                      % Diameter of insulated conductor of HV
    no_layers_hv=20;                     % Assume
    bp=no_layers_hv*dh_ins+(no_layers_hv-1)*0.3; % Radial Depth of HV
    lcp=(th/no_layers_hv)*dh_ins;        % Axial depth of HV
    Dih=Dol+2*ins_thick_hv;              %Inside diameter of HV
    Doh=Dih+2*bp;
    DH=(Dih+Doh)/2;                      % Mean diameter of HV

    % STEP7
    K=vphh/vph1;
    Lmtl=pi*DL*1e-3;                     % length of mean turn
    RL=0.021e-006*tl*Lmtl/AL;            % Resistance of LV winding
    LmtH= pi*DH/1000;                    % Mean length of HV winding
    RH=.021e-006*LmtH*th/ah;             % Resistance of HV winding
    R1=RL+RH/K;                          % Total resistance referred to LV side
    RP=R1*iph1/vph1;                     % pu resistance referred to LV side

    %STEP8
    %leakage reactance of transformer reffered to LV side(three phase core)%
```

```
lc = (lcs+lcp)/2;        %average height of winding%
a = .015;                %distance between HV &LV winding%
u0 = .000001256;
xl = 2*pi*f*u0*tl^2*Lmtl/lc*(a+((bp+bs)/3));
xlpu= xl*(iphl/vphl); % per unit leakage reactance
rlpu=RP;
puimpedance = sqrt((xlpu)*(xlpu)+(rlpu)*(rlpu)); %per unit impedance
%p.u. regulation%
cosx =0.8; %power factor%
sinx =sqrt(1-cosx^2);
puregulation = sqrt(rlpu*cosx+xlpu*a);


%step10
%Transformer losses
total_culoss=3*iphl*iphl*R1;        % Total copper losses
stray_losses=0.1 * total_culoss;    % Stray losses is 10 % of total Cu loss
Bmf=[0.7,0.9,1.1,1.3];
Byf=[0.6,0.75,0.92,1.08];
wmf=[0.83,1.33,2.0,2.83];
wyf=[0.66,1.16,1.5,1.92];
poly_wm = polyfit(Bmf,wmf,3);
poly_wy = polyfit(Byf,wyf,3);
wm=poly_wm(1)*bm^2 + poly_wm(2)*bm^1 + poly_wm(3);
wy=poly_wy(1)*by^2 + poly_wy(2)*by^1 + poly_wy(3);
dn_crgo_steel=7.6e3;
weight_core=ai*hw*dn_crgo_steel;
weight_yoke=ay*H*dn_crgo_steel;
iron_loss_core=wm*weight_core;
iron_loss_yoke=wy*weight_yoke;
iron_loss=iron_loss_core+iron_loss_yoke; % Total iron loss
Total_losses=iron_loss+ total_culoss + stray_losses;
% efficiency calculations
efficiency= kva*cosx/(kva*cosx+Total_losses);
% Calculations of AT of core
if bm <= 1.7
   atc= -7.6703* bm.^15+30.1849*bm.^14 -30.622*bm.^13 +15.3884*bm.^9 -
      15.9632*bm.^3+26.8724*bm-1.4687;
else
   atc= (0.4571* bm.^15-2.0683*bm.^14 +2.4958*bm.^13 -1.4753*bm.^10)*1000;
end;
% Calculations of AT of yoke
if by <= 1.7
   aty= -7.6703* by.^15+30.1849*by.^14 -30.622*by.^13 +15.3884*by.^9 -
      15.9632*by.^3+26.8724*by-1.4687;
else
   aty= (0.4571* hy.^15-2.0683*by.^14 +2.4958*by.^13 -1.4753*by.^10)*1000;
end;
```

```
total_mag_mmf=hw*atc+hy*aty;
if phase==3
   mag_mmf_per_phase=total_mag_mmf/3;
else
   mag_mmf_per_phase=total_mag_mmf;
end
Im=mag_mmf_per_phase/(2*tl); % Magnetizing current
Il=iron_loss/(3*vphl);   % Loss component of no load current
Io=sqrt (Im^2+Il^2);     % No load current

% step 13

% height of tank ht
ht=(2*by+hw)+50+150+ 200;

% width of tank wt=w+diameter of winding+clearance for tank valve on both sides
c_f_t_v_on_bs=54;
wt=W+ Doh+c_f_t_v_on_bs;

% length of tank=diameter of out winding d4+clearance for tapping wires on each
   side
cl_tw_on_eachside=33;
ln_tank=Doh+cl_tw_on_eachside;

% tubes are provided for cooling of transformer permissible temp rise of oil=Q
q=50;
c_trans_per_temprise=q;

% total area of four walls of tank=st
st=4*ht*wt*1e-6;
t_area_fourwalls_tank=st;
% let area of tubes = x*st
% x=total_loss/(st*q)-12.5)/8.8
x=(Total_losses/(st*q)-12.5)/8.8;

% total area of the tubes at= (total loss/q-12.5*st)/8.8
at=(Total_losses/q-12.5*st)/8.8;

% number of tubes n1= at/3.14*dt*it
dt=50;
lt=40;
n1=at/(3.14*dt*lt);

%%%Design Sheet%%%%%
disp(' ' )
disp(' ')
disp(' ')
disp({'Window width='ww 'meter'})
disp({'Height of window='hw 'meter'})
disp({'Total Frame width='w 'meter'})
disp({'Total Frame Height='H 'meter'})
disp({'Height of Yoke='hy 'meter'})
disp({'Depth of Yoke='a 'meter'})
```

```
disp({'Width of LV cond.=' Y ' meter'})
disp({'Height of LV cond.=' X 'meter'})
disp({'No. of Turns of LV=' tl })
disp({'No. of Turns of HV=' th })
disp({'Resistance of LV=' RL 'ohm' })
disp({'Resistance of HV=' RH 'ohm' })
disp({'PU Res. (LV)=' RP 'ohm' })
disp({'PU React.(LV)=' xlpu 'ohm' })
disp({'PU Imp. (LV)=' puimpedance 'ohm' })
disp({'PU Regulation=' puregulation 'ohm' })
disp({'Iron losses Wi=' iron_loss 'watts' })
disp({'Copper losses Wc='total_culoss 'watts' })
disp({'Total losses W='Total_losses 'watts' })
disp({'Efficiency =' efficiency '%' })
disp({'No load current=' Io 'Ampere'})
disp({'Height of the Tank=' ht 'meter' })
disp({'Weight of the Tank=' wt 'meter' })
disp({'Height of the Tank=' ln_tank 'meter' })
disp({'Number of Tubes =' nl })
```

2.2.14 Quiz on Transformer

1. What do you know about parallel operation of any electrical machine?
2. Why parallel operation of transformer is necessary in power systems?
3. What are the conditions for parallel operation of transformers?
4. Why is it preferable to install two or more transformers in parallel than one large unit?
5. Can you change the secondary terminal voltage by tape changer in the case of parallel operating transformers?
6. What will happen if the two transformers are not connected with proper polarity?
7. How can you reduce the losses in the power system by using the two transformers operating in parallel?
8. What will happen if dc supply is given to a transformer?
9. What will happen if the iron core of transformer is replaced by air core?
10. Why does the efficiency of transformer lie between 90% to 100%, and the efficiency of electrical motor or generator is always less than this?
11. What will be the effect of increasing number of turns of a coil?
12. How does operating frequency affect the transformer core loss?
13. If cosinusoidal supply is given instead of sinusoidal supply, will the transformer work or not? If yes then what will be the output?
14. On which factor the induced emf depends in the transformer secondary?
15. In a transformer, core flux depends on voltage, whereas leakage flux depends on current. Why?
16. What is the effect on the air core transformer, if an aluminum sheet is inserted in the secondary winding?
17. Draw the equivalent circuit of an air core transformer.
18. What is the main cause of noise in transformers and how it can be reduced?
19. While separating core loss components by variable frequency method, why do you keep voltage to frequency ratio constant?

20. What do you mean by the circulating current in transformers when they are connected in parallel?
21. What are the bed effects of circulating current on the parallel operation of the transformer at no load?
22. Discuss the effect if two transformers operating in parallel have different per unit impedances?
23. How is the parallel operation of the transformers economical for the power system in which there is variation of the load in a wide range?
24. Can you connect the transformers of different ratings for parallel operation? If yes then how?
25. How can you reduce the core losses in the transformer?
26. What do you mean by voltage regulation of transformer?
27. What happens to the secondary voltage when the transformer is loaded by leading power factor load?
28. Why core losses in the short circuit test and copper loss in open circuit test are negligible?
29. When transformer is loaded by same load with lagging power factor in one case and leading in the other. Is there any difference in the transformer efficiency in these cases?
30. What is the maximum efficiency condition of transformer?
31. Why does distribution transformer have low core losses while power transformer is having low copper loss?
32. Is it possible that regulation be negative, if yes then in which case?
33. At what power factor of the load the regulation will be zero?
34. Why is ac resistance of the winding more than the dc resistance?
35. Why are air core transformers used for high frequency applications, and iron core used for power frequency applications?

2.2.15 Review Exercises

I. Answer in Brief :

1. *Define* : (i) MMF, (ii) Reluctance, (iii) Permeability.
2. Mention two advantages of delta-delta connection of transformers.
3. Why delta-star transformer is used for step down purposes at the distribution substation.
4. State the essential conditions of parallel operation of single phase transformers.
5. State four applications of auto-transformer.
6. Why it is preferred to install two or more transformer in parallel rather than one large unit. Give four reasons.

II. Multiple Choice Questions :

1. An ac voltage of 220 V, 50 Hz is applied across a coil of 100 turns wound on iron core. The coil turns are now halved/doubled. The applied voltage in each case for the same core flux density is:

Turns halved	Turns doubled
(a) 440 V	110 V
(b) 110 V	440 V
(c) 220 V	220 V
(d) 110 V	110 V

2. A coil wound on a magnetic core is excited from an ac voltage source. The source voltage and its frequency are both doubled. The eddy current loss in the core will become:
   (a) remains same   (b) double
   (c) four times   (d) none of these

3. No load current in a transformer:
    (a) Lags the applied voltage by 90 deg.
    (b) Lags the applied voltage by somewhat less than 90 deg.
    (c) Leads the applied voltage by 90 deg.
    (d) Leads the applied voltage by somewhat less than 90 deg.

4. Two transformer connected in parallel shares load in the ratio of their kVA ratings only if their p.u. impedance (on their own kVA's) are :
    (a) equal
    (b) in the inverse ratio of their ratings
    (c) in the direct ratio of their ratings
    (d) purely reactive

5. Under balanced load conditions, the main transformer rating in the Scott connection is greater than that of the teaser transformer by:
    (a) 5%   (b) 15%   (c) 57.7%   (d) 85%

6. The voltage applied to a transformer primary is increased so as to keep V/f fixed. How will the core loss and magnetizing current change?

Core loss	**Magnetizing current**
(a) will increase	will remain same
(b) will remain same	will remain same
(c) will decrease	will increase
(d) will remain same	will decrease

7. When a bank of two single phase transformers in an open delta arrangement is used, each of them supplies :
    (a) 33.3% of its output rating
    (b) 86. 6 % of its output rating
    (c) 48.6 % of its output rating
    (d) 100 % of its output rating

8. Two single-phase transformers with proper connection can be used to achieve a three-phase output from three phase input :
    (a) True
    (b) False

9. The power transformer is :
    (a) Constant voltage device
    (b) Constant main flux device
    (c) Constant current device
    (d) Constant power device

10. Under balanced load condition, main transformer rating in Scott connection is :
    (a) 10 % greater than teaser
    (b) 15 % greater than teaser
    (c) 57.7 % greater than teaser
    (d) 66.6 % greater than teaser

11. What kind of bushing will be used in transformer above 33 KV ratings :
    (a) Porcelain type
    (b) condenser type
    (c) Oil filled type
    (d) (b) or (c)

12. The circuit of the autotransformer is shown in figure. The point b is located half way between a and c. The resistance of entire winding is 0.1 ohm and the resistance of bc is 0.04 ohm. what will be the copper loss at an output of 10 A. When the exciting current is zero :
    (a) 20 Watt
    (b) 2.5 watt
    (c) 10 W
    (d) 4 W.

13. Current phasor of core loss component of a transformer at no-load is
    (a) 90 deg. lag with voltage
    (b) in-phase with voltage
    (c) 90 deg. Lead to voltage
    (d) none of these

14. The open circuit test of a transformer gives :
    (a) Hysterisis loss   (b) Eddy current loss
    (c) sum of hysterisis and eddy current loss (d) Copper loss
15. Which test on a transformer provides information about regulation, efficiency and heating under load condition :
    (a) Open circuit test   (b) Back to Back test
    (c) Hopkinson test   (d) Short circuit test
16. Which loss varies significantly with load :
    (a) Hysterisis loss   (b) Eddy current loss
    (c) Copper loss   (d) Core loss
17. Three to three phase transformer connections not feasible for parallel operation is :
    (a) Y-Y to Y-Y   (b) Y-Y to delta-delta
    (c) Y-delta to delta -delta (d) Y-delta to delta-Y
18. Which of the following part is likely to suffer maximum damage due to excessive temperature rise.
    (a) Core laminations   (b) Copper winding
    (c) Dielectric strength of oil (d) Winding insulation
19. A distribution transformer usually a :
    (a) Star-delta transformer   (b) delta-star transformer
    (c) Star-star transformer   (d) delta-delta transformer.
20. Two transformers are operating in parallel will share the load depending upon their
    (a) Rating   (b) Leakage reactance
    (c) Efficiency   (d) per unit impedance.
21. A good transformer must have regulations as high as possible :
    (a) True   (b) False.
22. A transformer transforms :
    (a) Energy   (b) Frequency   (c) Voltage   (d) All of the these.
23. Which transformer will have smallest core :
    (a) 1 kVA, 50 Hz   (b) 1 kVA, 100 Hz   (c) 1 kVA, 200 Hz   (d) 1 kVA, 400 Hz.

III. Fill in the Blanks :

1. The turn ratio of teaser transformer is .... the main transformer in Scott connection.
2. The turn ratio of one transformer is .... the other transformer in open delta connection.
3. Transformation ratio of two winding transformer is .... when connected as an auto-transformer than when used as a two winding transformer.
4. kVA rating of two winding transformer is .... when connected as an auto-transformer than when used as a two winding transformer.
5. Open delta connection has a kVA rating of .... % of the rating of the normal delta-delta connection.
6. No-load current in transformer consisting of two components : (i) .... and (ii) ....
7. Eddy current losses are .... heating losses in the .... of the transformer.
8. The main reason for using an auto-transformer is ....
9. The no load current of transformer is .... percent of full load current.
10. An auto-transformer finds its applications extensively as ....
11. At lower frequencies .... laminations can be used.

12. The transformer tank is usually made of ....
13. A transformer has no .... losses.
14. The negative regulation is due to ....
15. The main disadvantage of auto transformer is the ....

Answers (transformer)

II. Multiple Choice Questions

1. (*b*)	**2.** (*c*)	**3.** (*b*)	**4.** (*b*)	**5.** (*b*)
6. (*a*)	**7.** (*b*)	**8.** (*a*)	**9.** (*b*)	**10.** (*c*)
11. (*d*)	**12.** (*b*)	**13.** (*b*)	**14.** (*c*)	**15.** (*b*)
16. (*c*)	**17.** (*c*)	**18.** (*d*)	**19.** (*b*)	**20.** (*d*)
21. (*b*)	**22.** (*c*)	**23.** (*d*)		

III. Fill in the Blanks

1. Less than (86.6%)	**2.** Equal to **3.** More	**4.** Greater
5. 58%	**6.** Iron loss, magnetizing	
7. Resistive, core	**8.** Copper saving	
9. 5	**10.** Variable voltage device	
11. Thicker	**12.** Mild steel	
13. Mechanical	**14.** Capacitive loads.	

15. No electrical isolation between primary and secondary windings.

SECTION–B

2.3   DIRECT CURRENT MACHINES

The d.c. machines were the first electrical machines invented. An elementary motor drove an electrical locomotive in Edinburgh in 1839 although it took another forty years before the d.c. motor really became widespread. It is still used to power trains and cranes.

FIGURE 2.45. *D.C. Motor*

2.3.1   Introduction

Figure 2.46 shows the general arrangement of a four pole d.c. motor or generator. The fixed part consists of steel core referred as pole cores, attached to a steel ring called yoke. The pole cores are made by silicon steel laminated plates riveted together and bolted to the yoke. Yoke may be of cast steel or fabricated rolled steel. Each pole core has a pole shoe, partly to support the field windings and partly to increase the cross sectional area and thus reduce the reluctance of the air gap. Each pole core carries a winding to make N and S poles of the field magnet. The Stator of a two pole d.c. machine is shown in Figure 2.47.

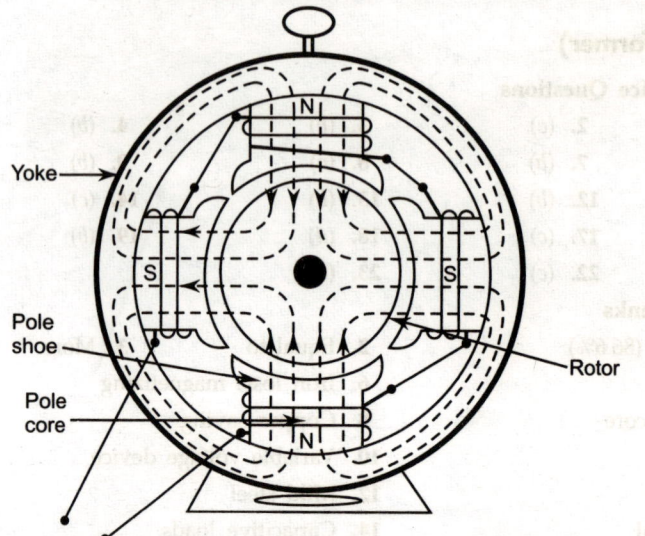

FIGURE 2.46. *General arrangement of d.c. machine*

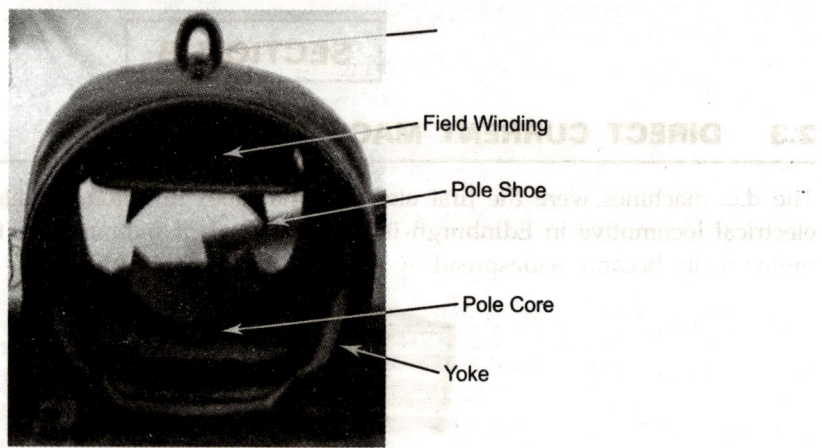

FIGURE 2.47. *Stator of a two pole dc machine*

The armature core consists of silicon steel lamination, about 0.4–0.6 mm thick, insulated from one another and assembled on the shaft in the case of small machines and on a cast steel spider in the case of large rating machines. The purpose of laminating the core is to reduce eddy current loss. Slots are stamped on the periphery of the laminations to keep armature winding in it as shown in Figure 2.48.

The dotted lines represent the distribution of useful flux into the armature core and link the armature conductors. The flux coming out from the north pole of a 4-pole machine divides into two paths, half is going into each S pole of a 4-pole machine. The armature MMF waveform when only armature winding is excited and field MMF waveform when only field winding is excited are shown in Figure 2.50(a). Suppose the armature rotates in clock wise direction, the emf

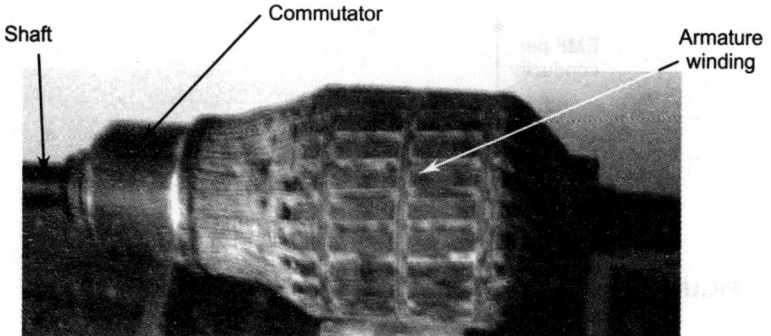

FIGURE 2.48. *Rotor of a two pole dc machine*

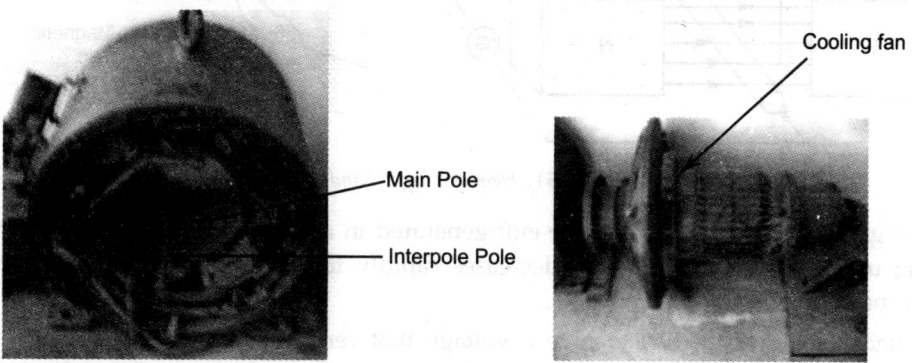

FIGURE 2.49. *Stator and rotor of a 4-pole dc machine*

is induced in armature coils as shown in Figure 2.50(b). The direction of emf may be determined from Fleming's right hand rule. This rule may be stated as :

"Stretch out the **forefinger, middle finger** and **thumb** of right hand so that they are right angles to one another. If the forefinger points the direction of magnetic field, thumb in the direction of motion of the conductor, then the middle finger will point the direction induced current." (Ref. Figure 2.51).

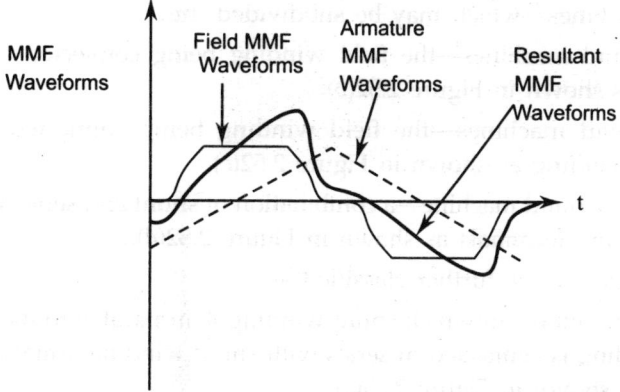

FIGURE 2.50. (*a*) *Waveform of EMF generated in armature winding*

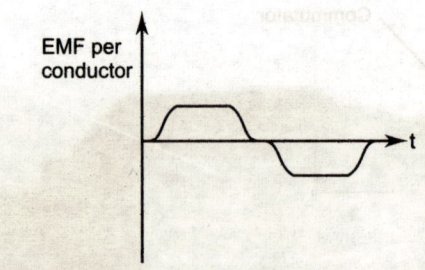

FIGURE 2.50. (*b*) *Waveform of EMF generated in armature winding*

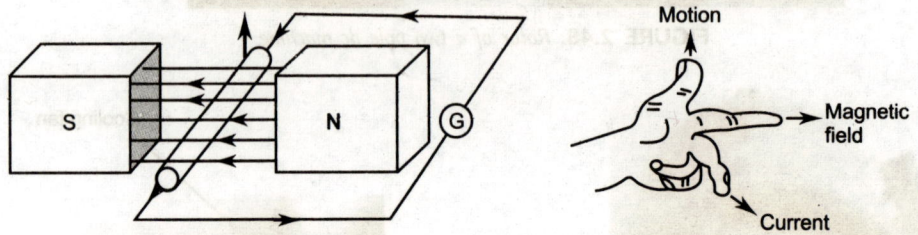

FIGURE 2.51. *Fleming's right hand rule*

If the air gap is of uniform length, the emf generated in a conductor remains constant while it is moving under pole face, and then decreases rapidly to zero when conductor is midway between the pole tips of adjacent poles.

A d.c. machine, however, has to give a voltage that remains constant in direction and in magnitude, and it is therefore necessary to use a commutator to enable a steady or direct voltage to be obtained from alternating emf.

2.3.2 Classifications

D.C. machine may be classified depending upon the connections of armature and field winding connections as follows (Ref. Figure 2.52) :

(*a*) Separately excited machines—the field winding being connected to a separate source of d.c. supply and not with the armature of d.c. machine as shown in Figure 2.52(*a*).

(*b*) Self excited machines—which may be subdivided into :

  (*i*) Shunt wound machines—the field winding being connected across the armature terminal as shown in Figure 2.52(*b*).

  (*ii*) Series wound machines—the field winding being connected in series with the armature winding as shown in Figure 2.52(*c*).

  (*iii*) Compound wound machine—a combination of shunt and series windings connected with armature terminals as shown in Figure 2.52(*d*).

Compound machines may be further classified as

(*a*) Short shunt compound—in which shunt winding is in parallel to the armature winding. The series winding is connected in series with shunt winding-armature winding parallel combination as shown in Figure 2.53(*a*).

(*b*) Long shunt compound—in which shunt winding is in parallel to both the armature winding and series winding as shown in Figure 2.53(*b*).

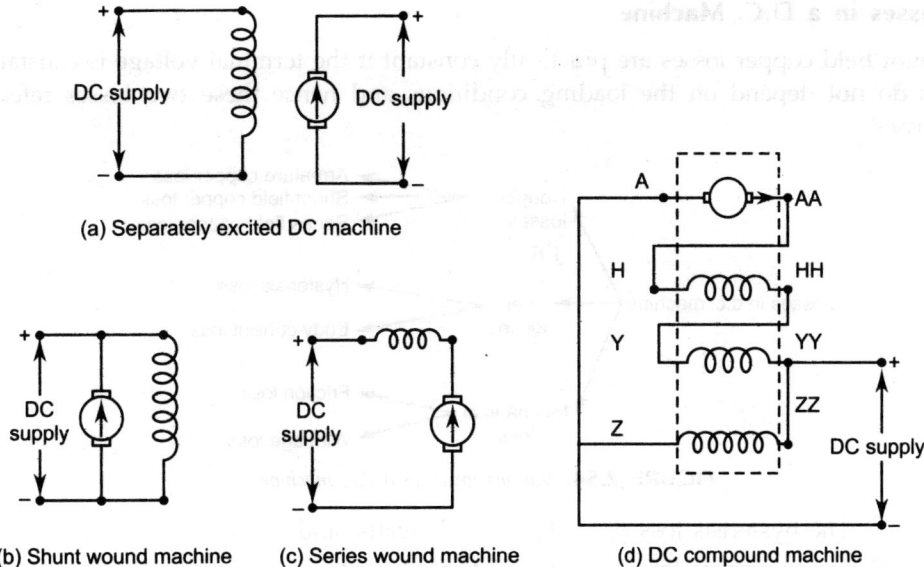

FIGURE 2.52. *Different types of separately and self excited d.c. machines*

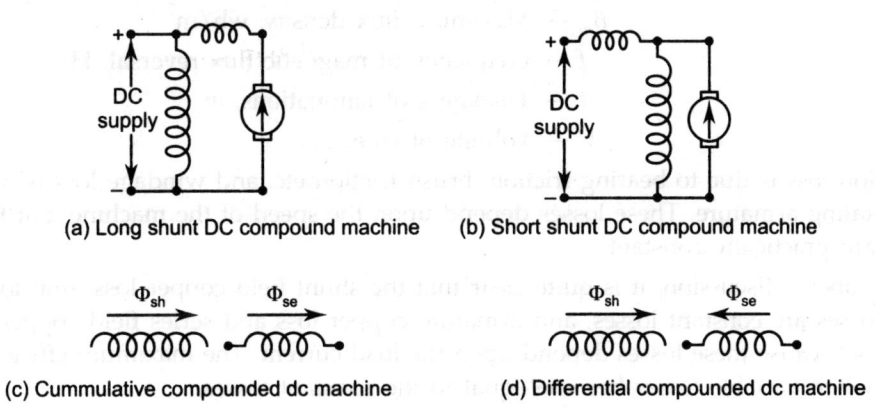

FIGURE 2.53. *Different types DC Compound Machines*

Compound machines may also be categorized depending on the direction of flux in series and shunt windings:

(a) Cumulative compounded machine—In these types of machines the series winding flux supports the shunt winding flux and the net air gap flux is the sum of series flux and shunt flux as shown in Figure 2.53(c).

(b) Differential compounded machine—the series field winding flux opposes the shunt field winding flux, therefore, net air gap flux is the difference of shunt field flux and series field flux as shown in Figure 2.53(d).

$$\phi_{net} = \phi_{sh} \pm \phi_e \text{ where + sign for cumulative compounding and}$$
$$- \text{sign for differential compounding.}$$

Based on the flux magnitudes the compound machines may also be classified as over compounded, under compounded or level compounded machines.

2.3.3 Losses in a D.C. Machine

The shunt field copper losses are practically constant if the terminal voltage is constant. The iron losses do not depend on the loading conditions and hence these two losses referred as constant losses.

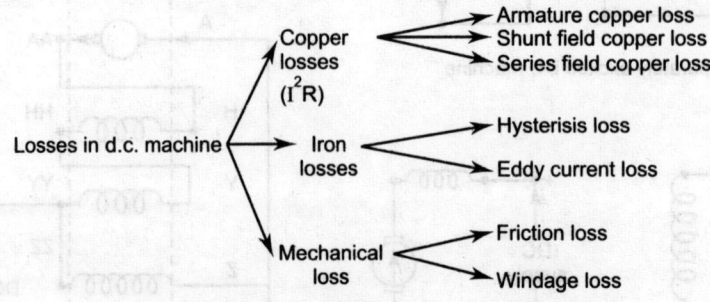

FIGURE 2.54. *Various losses in a d.c. machine*

The hysterisis loss $P_h = K_h B_m^{1.6} f V$ watts, and

The eddy current loss $P_e = K_e B_m^2 f^2 t^2 V$ watts

Where K_h or K_e are constants

B_m — Maximum flux density, wb/m^2

f — Frequency of magnetic flux reversal, Hz

t — Thickness of laminations, m

V — Volume of core, m^3.

The friction loss is due to bearing friction, brush friction etc. and windage loss is due to air friction of rotating armature. These losses depend upon the speed of the machine, but for given speed, they are practically constant.

From the above discussion, it is quite clear that the shunt field copper loss, iron losses and mechanical losses are constant losses, and armature copper loss and series field copper loss are variable losses because these losses depend upon the load current. The maximum efficiency may be obtained when variable losses become equal to the constant losses.

2.3.4 Characteristics of D.C. Machines

(i) D.C. Shunt Generator

(a) *Open circuit characteristics* : The open circuit characteristic of a shunt generator is similar to that of series generator as shown in Figure 2.55.

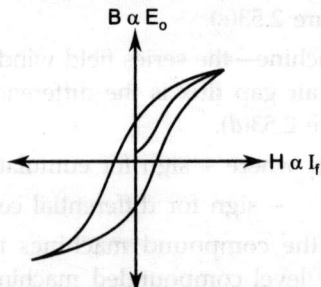

FIGURE 2.55. *Hysterisis loop or open circuit characteristic*

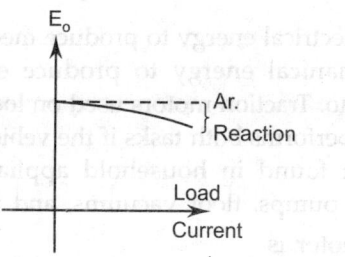

FIGURE 2.56. *Internal characteristic*

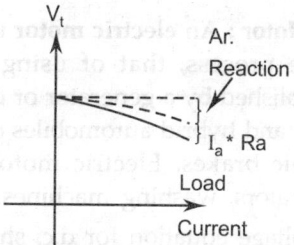

FIGURE 2.57. *External characteristic*

The area under hysterisis loop is called hysterisis loss, which may be reduced by using better magnetic material like cold rolled grain oriented silicon steel.

(b) *Internal characteristics :* When the generator is loaded, the flux per pole is reduced due to armature reaction. Therefore the generated emf on load is less than the emf generated at no load as shown in Figure 2.56.

(c) *External characteristics :* It gives the relationship between terminal voltage and the load current as shown in Figure 2.57. The equation for terminal voltage in case of shunt generator is

$$V_t = E_o - I_a * R_a$$

(ii) **D.C. Series Generator :** In series generator as the load current increases the series flux increases, which in turn increase the generated emf as shown in Figure 2.58. The equation for terminal voltage in case of series generator is

$$V_t = E_o - I_a (R_a + R_{se})$$

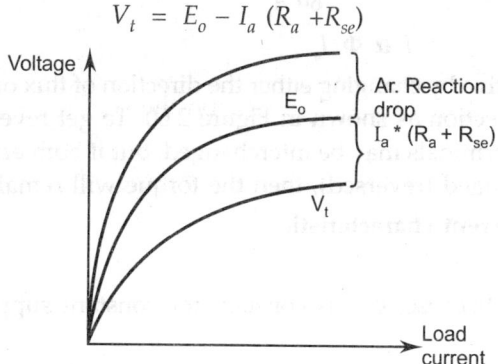

FIGURE 2.58. *Internal and External characteristic for series generator*

(iii) **D.C. Compound Generator**

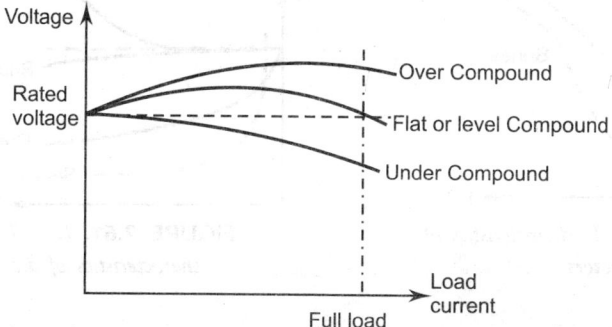

FIGURE 2.59. *Internal and External characteristic for compound generator*

(*iv*) **D.C. Motor :** An **electric motor** uses electrical energy to produce mechanical energy. The reverse process, that of using mechanical energy to produce electrical energy, is accomplished by a generator or dynamo. Traction motors used on locomotives and some electric and hybrid automobiles often performs both tasks if the vehicle is equipped with dynamic brakes. Electric motors are found in household appliances such as fans, refrigerators, washing machines, pool pumps, floor vacuums, and fan-forced ovens.

The voltage equation for d.c. shunt motor is

$$V_t = E_o + I_a R_a$$

Derivation of Torque Equation

The generated power in d.c. motor is

$$P = I_a E_o = T N$$

where

I_a = Armature current, ampere

E_o = back emf of armature, volts

T = Torque, N-m

N = speed, rpm

$$T = \frac{I_a E_o}{N} = \frac{I_a \phi ZNP}{60\,AN} = k\phi\,I_a$$

where

$$k = \frac{ZP}{60\,A}$$

$$T \propto \Phi\,I_a$$

The torque may be negative by changing either the direction of flux or armature current to rotate the motor in the reverse direction as shown in Figure 2.60. To get reverse speed, either armature terminals or field winding terminals may be interchanged, but if both armature and field terminals are simultaneously interchanged (reversed), then the torque will remain positive.

(*a*) **Torque-armature current characteristic**

Shunt motors

$T \propto I_a$ because, Φ – is constant for constant supply voltage.

Series motors

$T \propto I_a^2$ because, $\Phi \propto I_a$.

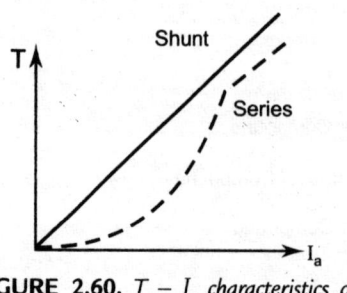

FIGURE 2.60. $T - I_a$ *characteristics of d.c. motors*

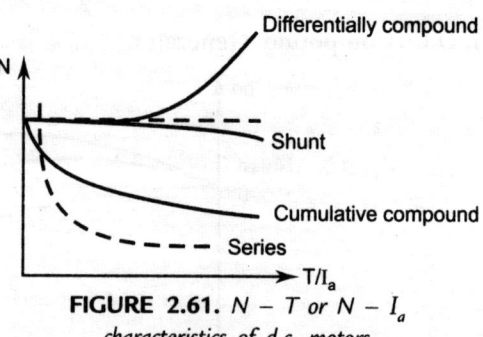

FIGURE 2.61. $N - T$ *or* $N - I_a$ *characteristics of d.c. motors*

(b) Speed current characteristics

$$E_o = V_t - I_a R_a \rightarrow \frac{\phi ZNP}{60\,A} = V_t - I_a R_a$$

$$N = \frac{V_t - I_a R_a}{K\phi} \rightarrow N \alpha \frac{E_0\,(= V_t - I_a R_a)}{\phi}$$

There is a slight change in the speed of shunt motor from no-load to full load, hence it is called drooping N-T characteristics d.c. motor. On the other hand, in series motors the speed automatically changes as load changes. Thus, if load decreases, its speed is automatically raised and vice versa. Hence, this characteristic is called **self releasing** characteristic of d.c. series motors as shown in Figure 2.61. At no load the, the armature current is very small and so is the flux. Hence, the speed of series motor is excessively high at no load, because $N \alpha \dfrac{1}{\phi} \alpha \dfrac{1}{I_a}$. It may damage the motor due to centrifugal force set up in rotating parts. Therefore, series motors should never be started at no load.

2.3.5 Comparison between Inter Poles and Compensating Winding

Inter Pole Winding

These are small poles placed midway between main poles directly over conductors being commutated. Inter pole flux cancels out armature reaction flux and eliminated sparking at brushes. It has no influence on flux weakening in main field poles due to armature reaction. Their influence does not extend that far.

Compensating Windings

They are used in dc machines with heavy duty cycle. The goal is to completely cancel armature reaction and eliminate shifting of magnetic neutral plane and also main pole flux weakening. These are the windings housed in slots carved in pole faces as shown in Figure 2.62 and connected in series with the armature.

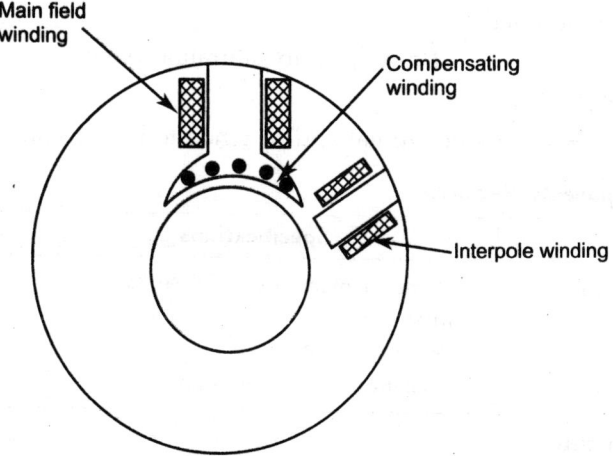

FIGURE 2.62. *Compensating winding and interpole winding in a dc machine*

2.4   EXPERIMENTS ON D.C. MACHINES

2.4.1   To Determine the Magnetization Characteristics

To determine the magnetization characteristics or open circuit characteristic (OCC) and the speed verses voltage curve of a separately excited D.C. Generator.

Necessity of the Experiment

It is found that there are many advantages of ac systems, but then also the dc machines are used in industries because

    (i) D.C. machines are highly flexible and versatile energy conversion devices.

    (ii) D.C. machines possess high starting torque.

    (iii) A typical D.C. machine can meet the high accelerating and decelerating torques.

    (iv) D.C. drives are useful where wide range of speed control is required.

    Due to above features the D.C machines are used in industries, although they have higher initial cost. Hence, it is necessary to know about the performance characteristics of D.C. machines.

Theory

No load characteristic gives the variation of armature generated e.m.f. E_a with field current I_f at zero armature current and constant speed. It will be seen that even though the field winding is not energized, the voltmeter indicates a small voltage (2 to 6 volts) due to presence of residual flux in the main poles. Then connect field circuit and draw O.C.C. by changing field current I_f by changing field excitation. It is seen that the characteristics is not reversible when I_f decreased, then O.C.C. will not follow original curve but will lie above it due to hysterisis.

$$E_a = \frac{\phi ZNP}{60 A}$$

At constant speed

$$E_a \; \alpha \; \Phi_f \text{ and}$$
$$\Phi_f \; \alpha \; I_f$$

When circuit is unsaturated,

$$Ea \; \alpha \; I_f \text{ up to saturation point}$$

After saturation $\Phi_f \neq K I_f$

Curve becomes more or less horizontal and E_a tends to be constant.

Equipments and Components Required

S.No.	Name	Specifications	No. Required
1.	D.C. Motor	Name of manufacturer BHP/KW ... RPM ..... Voltage ...... A......... Insulation ..... Connection	1
2.	D.C. Ammeter		2
3.	D.C. Voltmeter		2
4.	Rheostats		1

Connection Diagram

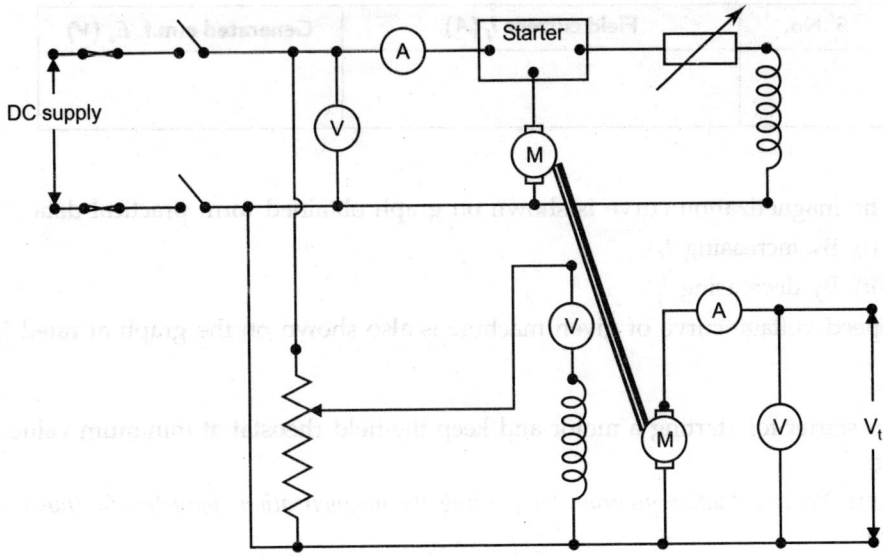

FIGURE 2.63. (*a*) *Connection diagram for OCC of D.C. machine*

Procedure

1. Make the connections as shown in Figure 2.63(*a*).

2. Start the D.C. shunt motor using starter and bring it to rated speed.

3. Note down the reading of voltmeter for E_a of the generator when field switch is open $I_f = 0$.

4. Close field Switch and increase I_f in steps and note down the corresponding terminal voltage E_a.

5. Precaution is to be taken that during increasing the field current I_f does not decrease it.

6. Field current I_f is increased till E_a is about 1.1 to 1.25 times the rated voltage.

7. Now decrease the field current in steps and note down the corresponding E_a.

8. While decreasing the field current I_f care is to be taken that it should not increase, otherwise the readings will be wrong.

9. Now set I_f at the rated value, vary the speed and note the terminal voltage of the generator keeping I_f constant.

Observations

(*i*) Speed Vs Voltage Curve at Rated I_f

S. No.	Speed N (r.p.m.)	Voltage E_a (V)

(*ii*) O.C.C. of separately excited generator rated speed = ... rpm

S. No.	Field current I_f (A)	Generated e.m.f. E_a (V)

Results

(*a*) The magnetization curve is shown on graph obtained form practical data
    (*i*) By increasing I_f
    (*ii*) By decreasing I_f

(*b*) Speed voltage curve of given machine is also shown on the graph at rated I_f.

Precaution

Use proper starter for starting a motor and keep the field rheostat at minimum value at the time of starting.

Problem. *Write a Matlab program for plotting the magnetization characteristic (hysterisis loop) for DC machine.*

```
% Matlab Program for Hysterisis loop
clear all; clc;
% DC machine field current and open circuit voltage
% If Voc
Readings=[0.1000  30.0000
0.2000  75.0000
0.3000  115.0000
0.4000  150.0000
0.5000  175.0000
0.6000  195.0000
0.7 201
0.6 199
0.5000  185.0000
0.4000  160.0000
0.3000  130.0000
0.2000  90.0000
0.1000  50.0000
-0.1000  -30.0000
-0.2000  -70.0000
-0.3000  -105.0000
-0.4000  -135.0000
-0.5000  -170.0000
-0.6000  -195.0000
-0.7 -200
-0.6 -198
-0.5000  -182.0000
-0.4000  -165.0000
-0.3000  -135.0000
-0.2000  -100.0000
```

```
-0.1000  -70.0000
 0.1 05.0000
 0.2000 55.0000 .
 0.3000 95.0000
 0.4000  140.0000
 0.5000  170.0000
 0.6000  195.0000
 0.7 203];
plot(Readings(:,1), Readings(:,2))
xlabel('Field Current, Amp.')
ylabel('Open circuit Voltage, volts')
```

Result of above mentioned program is shown in Figure 2.63(*b*)

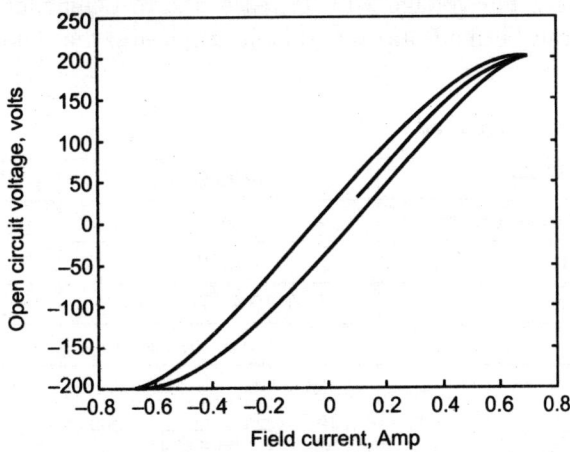

FIGURE 2.63. (*b*) *Hysterisis loop for DC machine*

EXPERIMENT NO. 6

2.4.2 To Perform the Load Test on D.C. Separately Excited Generator and Shunt Generator

Also draw the following characteristics and compare it.

(*i*) The external characteristics (V_t verses load current)

(*ii*) Load saturation curve (V_t verses I_f)

Necessity of the Experiment

Generally, it is seen that when the load on generator increases form no-load to full load, the terminal voltage does not remain same, but decreases somewhat, due to armature reaction and voltage drop in armature which are practically absent under no-load conditions. Hence, it is necessary to study the performance characteristics under load condition.

Theory

(*a*) **Load saturation curve (V_t verses I_f).** It is seen that at no-load the field ampere turns (AT_f) are required for rated no-load voltage.

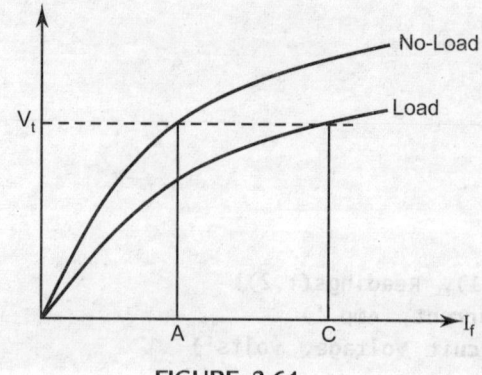

FIGURE 2.64

Under load conditions, the voltage will decrease due to demagnetizing effect of armature reaction. This decrease can be made up by suitable increasing the field current I_f as shown in Figure 2.64.

Equipments and Components Required

S. No.	Name	Specifications	No. required
1.	DC Motor generator set		1
2.	DC Voltmeter	0–250 V	2
3.	DC Ammeter	0–10 A, 0–2.5 A	2
4.	Rheostat	1000 ohm, 1 A	1

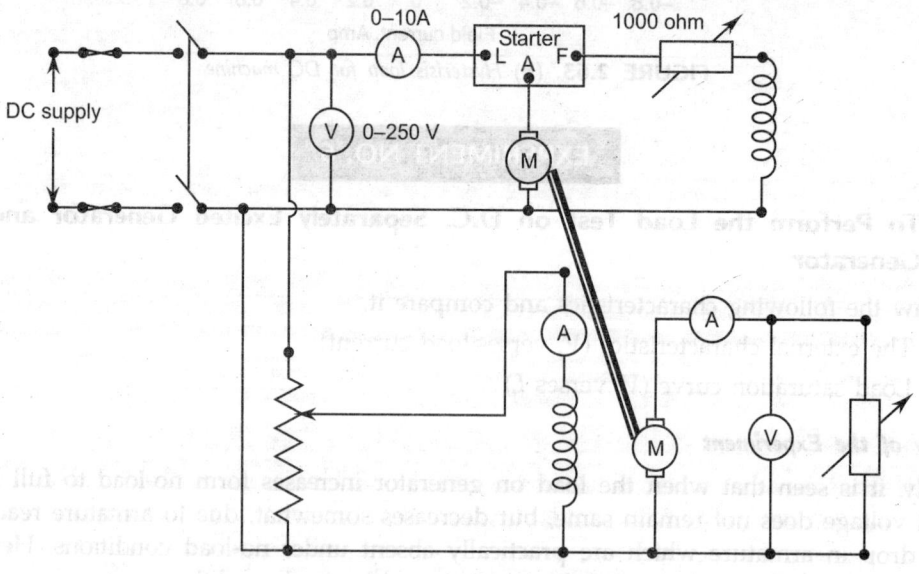

FIGURE 2.65. *Connection diagram for load Test of Separately D.C. generator*

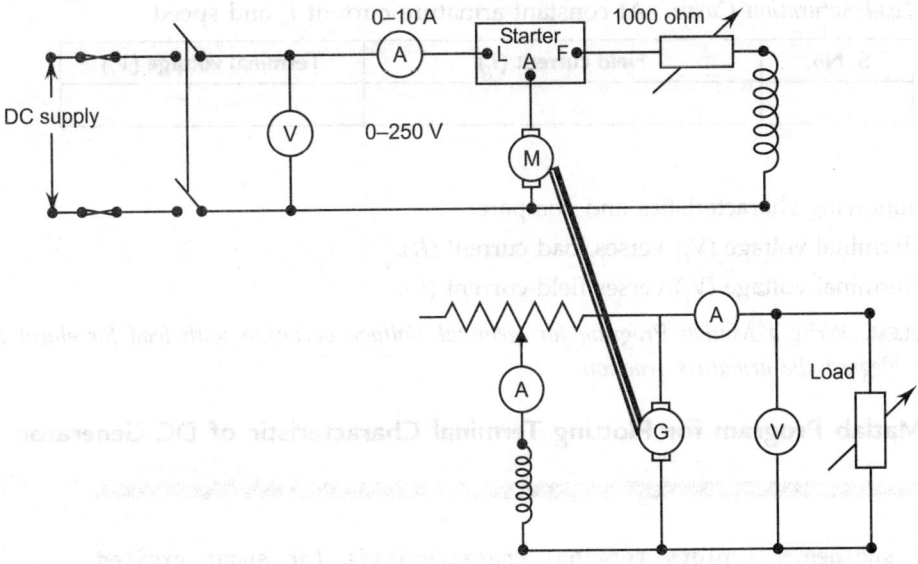

FIGURE 2.66. *Connection diagram for load Test of D.C. shunt generator*

(b) **Load Test on DC Shunt Generator**

Procedure

(a) *External Characteristic :*

1. Make the connections as show in Figure 2.65 for separately excited generator and after that connect as in Figure 2.66 for shunt generator.

2. Start motor and run at rated speed by adjusting field current.

3. Then increase the generator field current to obtain the rated voltage.

4. Keeping the field current and speed constant, take readings of terminal voltage V_t by increasing the load current.

5. Measure the armature and field winding resistances of the generator by the ammeter and voltmeter method.

(b) *Load Saturation Curve :*

6. After step 4, keep speed and armature current constant. Adjust I_f so that I_a equal to rated armature current. Take the readings of I_f and V_t.

7. Vary the load in steps and adjust I_f such that I_a remains constant but V_t changes. Note down the values of V_t and I_f at different loads.

Observations

(a) *External Characteristics :* At constant speed and field current I_f, take the following readings :

S. No.	Load current (I_L) A	Terminal voltage (V_t) (Volts)

(b) *Load Saturation Curve :* At constant armature current I_a and speed

S. No.	Field current (I_f)	Terminal voltage (V_t)

Results

Plot the following characteristics and compare:

(i) Terminal voltage (V_t) verses load current (I_L).

(ii) Terminal voltage (V_t) verses field current (I_f).

PROBLEM. *Write a Matlab Program for terminal Voltage variation with load for shunt excited dc generator. Neglect the armature reaction.*

2.4.3 Matlab Program for Plotting Terminal Characteristic of DC Generator

```
%%%%%%%%%%%%%%%%%%%%%%%%%%%%%%%%%%%%%%%%%%%%%%%%%%%%%%%%%%%%%%%
%
% shntgen.m - plots terminal characteristic for shunt excited
%    dc generator. Neglecting Armature reaction.
%
%%%%%%%%%%%%%%%%%%%%%%%%%%%%%%%%%%%%%%%%%%%%%%%%%%%%%%%%%%%%%%%
clear all; clf; clc;
VtR=230;         % Rated terminal voltage
PR=10000;      % Rated output power
Ra=0.01;         % Armature resistance
Rf=100;        % Field resistance
ILR=PR/VtR;   % Rated output current
npts=300;
IL=linspace(0,1.5*ILR,npts);
If1=VtR/Rf;If=If1;
E=VtR;
% Iterative determination of Vt-IL
for i=1:npts
    Vt1=E -(IL(i)+If1)*Ra
    if Vt1<0; break; end;
    x(i,:)=[IL(i), Vt1, E];
    If2=Vt1/Rf;
    E=VtR*If2/If
    If1=If2;
end
%axis([min(IL),max(IL),0,E]);
plot(x(:,1),x(:,2),'-')
hold on ;
plot(x(:,1),x(:,3),'-'); grid;

title('Shunt excited dc generator');
xlabel('Load current, A'); ylabel('Terminal voltage, V');
%*****************************************************************
```

PROGRAM OUTPUT is shown in Figure 2.67.

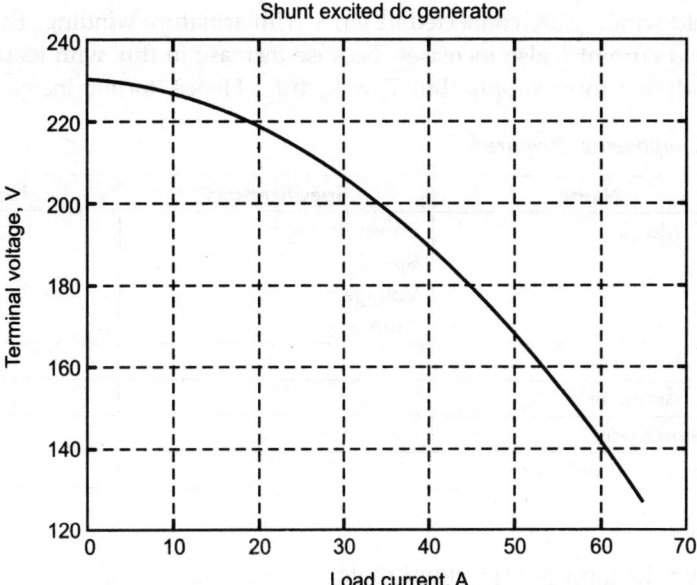

FIGURE 2.67. *Drop in terminal voltage of DC shunt generator.*

EXPERIMENT NO. 7

2.4.4   To Perform the Load Test on Series and Shunt Motors and from Test Data Plot the Following :

1. The speed—torque characteristics
2. The torque—current characteristics.

Necessity of the Experiment

In the industries where smooth and wide range of speed control is required, The D.C. motors are best suited for such applications. Different types of motors are used for different type of loads. For example, high starting torque is needed; D.C. series motors are used, *e.g.*, in railways, cranes etc. D.C. shunt motors are used where constant speed is required. Hence, it is necessary to study the speed-torque characteristics of motors.

Theory

In case of D.C. shunt motors, the field current is constant, if the supply voltage is kept constant. At small value of I_a the demagnetizing effect of armature reaction is almost negligible and therefore the air gap flux is unaffected. But for large value of I_a the demagnetizing effect of armature reaction, decreases the air gap flux slightly.

$$\text{Speed of D.C. motor } N = \frac{E_a}{K_a \phi} = \frac{V_t - I_a R_a}{K_a \phi}$$

Φ—field flux which is idealy constant, but due to increase in I_a, armature drop increases, which in turn increase the demagnetizing effect of armature reaction. Due to which, E_a decreases, therefore, speed slightly reduces. D.C. shunt motors have drooping characteristics.

Series Motors

In these motors field winding are connected in series with armature winding. The armature current I_a increases, the field current I_f also increases. Because increase in flux with load, speed must drop and large current drawn form supply. But $T_e = K_a \Phi I_a$. Hence, torque increases with load.

Equipments and Components Required

S. No.	Name	Specifications	No. required
1.	D.C. Motor	Power Rating = Speed = Voltage = Current =	1
2.	D.C. Voltmeter		1
3.	D.C. Ammeter		1
4.	Techno Meter		1
5.	Suitable Motor Load		1

Connection Diagrams

(a) Connection diagram of D.C. shunt motor

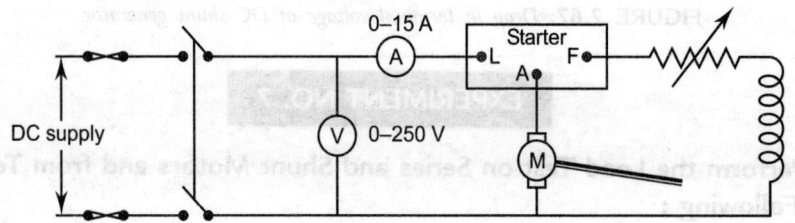

FIGURE 2.68. *Connection diagram for D.C. shunt motor*

(b) Connection diagram of D.C. series motor

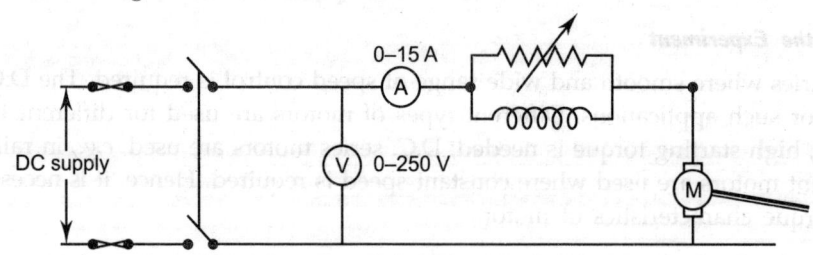

FIGURE 2.69. *Connection diagram for D.C. series motor*

Procedure

(a) *Load test on D.C. shunt motor*

    1. Connect the devices as in Figure 2.68.

    2. Run the motor on no-load and adjust its speed to the rated value.

    3. Increase the load torque or load on D.C. motor and note down the speed and corresponding armature current.

    4. Repeat the experiment so that I_a reaches to 120% of its rated value.

(b) *Load test on D.C. series motor*

1. Make the connections as shown in Figure 2.69.
2. Apply the load by tightening the belt or brake shoe appropriately.
3. Turn-on the main switch and start the motor through starter.
4. Adjust the brake shoe such that speed is 120% of its rated speed.
5. Note down the readings of ammeter I_a, voltmeter V_t and tachometer for speed.
6. Adjust the load on shaft in steps so that motor current reaches to 120% of its rated value.
7. Take the readings.

Observations

(a) D.C. Shunt motor

S. No.	I_a	V_t	Speed	Field current	Torque $I_a * I_f$

(b) D.C. series motor

S. No.	I_a	T_L (α I_a^2)	Speed	V_t

Result

The conclusions drawn from the graphs are depicted as follows:

1. In D.C. shunt motors speed is almost constant with load torque under normal operating conditions. The torque varies linear with I_a.

2. In D.C series motors speed varies with load when load is small then motor runs at higher speed than the speed at larger loads. The torque $-I_a$ curve drawn is as shown in the Figure 2.70.

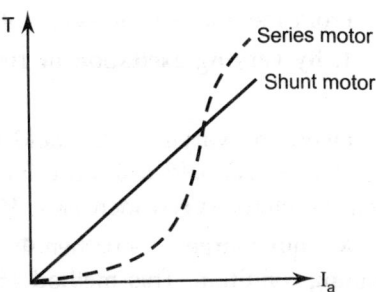

FIGURE 2.70. *Torque—I_a characteristics*

NOTE ▼ If it is not possible to measure torque accurately then measure armature current and field current and take their product to measure the Torque, *i.e.*, $T = K \Phi I_a = K I_f I_a$.

Precaution

Never start or run D.C. series motor at no-load or light loads, because at no-load or light loads the motor speed is large enough to damage the motor parts.

2.4.5 To Control the Speed of D.C. Shunt Motor

1. By varying the field excitation
2. By varying the armature resistance
3. By varying the armature terminal voltage (Ward Leonard control device)

Also draw the speed torque curves for different methods and compared

Necessity

D.C. Motor is generally used to drive the loads where variable speeds are required because of :

(a) Wide range of speed control

(b) Smooth control

(c) Easy and efficient methods to control the speed

Hence, it is necessary to have some idea about various speed control methods.

Theory

The equations governing to the D.C. shunt motor as given below :

$$E = V_t - I_a R_a \qquad \qquad ...(1)$$

and

$$E = \frac{\Phi ZNP}{60 A} \qquad \qquad ...(2)$$

$$E = K_a \Phi N$$

where

$$K_a = \frac{ZP}{60A}$$

Hence,

$$K_a \Phi N = V_t - I_a R_a \quad \Rightarrow \quad N = \frac{V_t - I_a R_a}{K_a \Phi} \qquad \qquad ...(3)$$

From the above expression different speed control methods can be as follows :

1. By varying excitation or field flux Φ : The field flux is directly proportional to I_f.

$$\Phi \ \alpha \ I_f$$

Hence, by varying I_f the field flux can vary. It is possible by inserting a series resistance in field circuit. As field resistance increase, I_f decrease which in turn decreases Φ. If flux decreases then the motor speed increases. The highest speed is limited by armature reaction.

R = quite large, $I_f \to 0$, then $\Phi \to 0$ and speed is infinity. So never open the field circuit under running condition. This method of speed control is called constant power drive.

2. By varying armature resistance : If armature resistance varies, which is inserted in series with armature circuit, the armature current changes as follows :

$$I_a = \frac{V - E}{r_a} = \frac{V - K_a \Phi N}{r_a} \qquad \text{when no resistance in armature circuit.}$$

$$I_{a2} = \frac{V - K_a \phi N}{r_a + R_g} = I_{a1} \frac{r_a}{R_g + r_a} \quad \text{when resistance } R_g \text{ inserted in armature circuit current decreases.}$$

Torque $T = K_a \Phi I_a$ which also decreases with current.

For constant load torque and generated torques T decreases then the speed decreases, which, in turn E decreases. As a result of it I_a decreases under steady state

$$N_1 = \frac{E_1}{r_a}, \; N_2 = \frac{E_2}{r_a + R_g}$$

For fixed value of resistance, the speed varies widely with load, since the speed depends on the voltage drop in this resistance.

Disadvantages of this method :

(*i*) Low efficiency and higher operational cost at reduced speed

(*ii*) Poor speed regulation

Only advantage of this method is that very low speed (a few r.p.m) can be obtained easily.

3. By varying armature terminal voltage : As V_t varies, counter e.m.f. $E = V_t - I_a R_a$ changes almost proportionally and for constant flux motors the speed changes in the same proportion as V_t .

The speed control by varying the armature terminal voltage is obtained by

(*a*) Ward Leonard System

(*b*) Controlled rectifiers.

Equipments and Components Required

S. No.	Name	Specifications	No. required
1.	Test Motor		1
2.	Motor generator set forward Leonard system		1
3.	Rheostat	1000 Ω, 1 A	1
4.	D.C. Voltmeter	0–250 V	2
5.	D.C. Ammeter	0–10 A, 0–2.5 A	1
6.	Mechanical Load		1
7.	Tachometer		1

Connection Diagram

(*a*) Connection diagram for speed control by armature resistance and field control method.

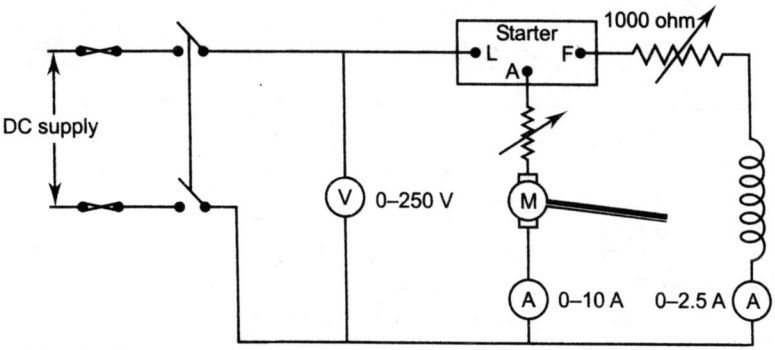

FIGURE 2.71. *Connection diagram for D.C. shunt motor for speed control*

(b) Schematic diagram of Ward Leonard system of speed control.

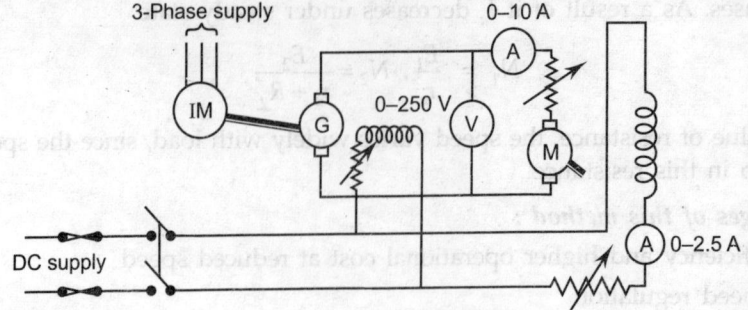

FIGURE 2.72. *Connection diagram of Ward Leonard System of speed control for D.C. motor*

Procedure

(a) *Speed control by field and armature resistance*

1. Make the connections as in Figure 2.71 and set external armature resistance at zero value.
2. Start the motor using the starter. Keep field current at its maximum value.
3. Gradually increase the field resistance and observe the variation of speed with different field currents under no-load condition.
4. Repeat the procedure for different constant loads.
5. Adjust the field resistance for rated field current and observe the variations in speed by changing Armature resistance for no-load and then for constant loads.

(b) *By varying armature terminal voltage*

1. Make connections as shown in Figure 2.72.
2. Start the three phase induction motor of the Motor-Generator (M-G) set.
3. Excite the field of the separately excited generator of the M-G set
4. Adjust the armature terminal voltage of test motor at rated value after exciting the field by cutting the resistance gradually.
5. Change the M-G set excitation currents and observe the speed of the motor on no-load.
6. Repeat the producer for different loads also.

Observations

(a) Field Flux Control :

S. No.	Armature current I_a (A)	Field current I_f (A)	Speed N (rpm)
1.	At No-load I_a		
2.			
3.			
4.			
5.	At load I_{a2}		
6.			
7.			
8.			
9.	At load I_{a3}		

(b) Armature Resistance Control :

At constant field current I_f =

S. No.	Field current I_f	Speed

(c) Ward Leonard Control :

(i) At No-load

S. No.	Field current of M-G set (A)	Test motor voltage V_t (volt)	Speed (rpm)

(ii) Constant torque operation

Constant I_{fm} = I_{am} =

S. No.	Speed (rpm)	Terminal voltage V_t (V)	Generator current I_{fg}	Field

Change the load on shaft.

Results

We have seen that motor speed varies as shown in Figure 2.73.

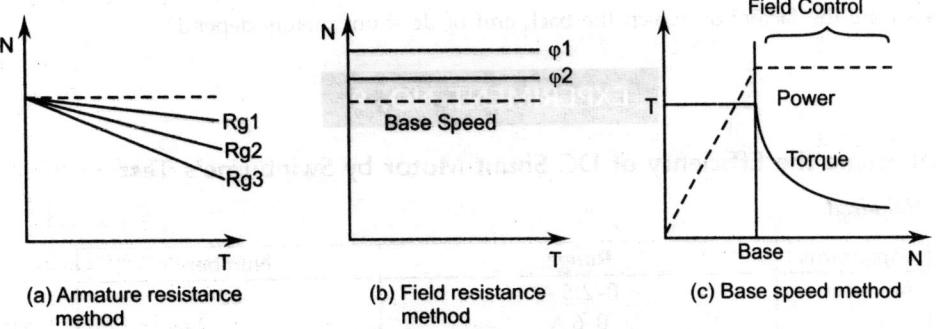

(a) Armature resistance method   (b) Field resistance method   (c) Base speed method

FIGURE 2.73. *Speed characteristics of dc shunt motor*

Draw the graphs between speed Vs torque and compare.

Precautions

1. If field opens suddenly due to loose connections or any other reason then armature speed becomes infinite and motor will get damaged because flux tends to reduce to a very small value.

$$N = \frac{V_t - I_a r_a}{\phi} \text{ when } \Phi = 0 \text{ , } N = \infty$$

2. Check the field connections of motor before starting of the motor.
3. Connections should be tight.

2.4.6 Experimental Quiz on D.C. Machine

1. What do you mean by the following terms?
   (a) Base speed,   (b) Speed regulation,
   (c) Constant power drive, and   (d) Constant torque drive.
2. Compare field flux control method and armature resistance control method.
3. List merits and demerits of field flux speed control method.
4. How does the dc motor rating depend on its speed range? Discuss in short.
5. What happens if the following situations arise when starter is connected or it is not connected in the circuit?
   (a) Field connections are reversed.
   (b) Supply terminals are reversed.
   (c) Field circuit opens in running condition.
   (d) Field circuits opens in starting condition.
   (e) Some field turns short circuited.
6. Which method is used for getting speeds below base speed? How?
7. Is it possible to obtain the speed twice the base speed? If yes, explain how?
8. How do speed torque characteristics change in the armature resistance control method?
9. In which speed control method the output torque remains constant? Explain.
10. For a dc motor the field flux speed control method is called constant power drive method. Justify your answer.
11. Name the methods by which you can control the speed of dc series motors.
12. Now-a-days the thyrister voltage control method is intensively used in practice for speed control.
13. What are the factors on which the back emf of dc shunt motors depend?

EXPERIMENT NO. 9

2.4.7 Determine the Efficiency of DC Shunt Motor by Swinburne's Test

Apparatus Required

Apparatus	Range	Number
Ammeter	0–2.5 A	02
	0–6 A	
Voltmeter	0–250 V	01
Rheostat	100 ohm, 5 A	02
	360 ohm, 1 A	

Theory

There are various losses in dc machines and may be listed as:

1. Iron losses

    (*a*) Hysterisis loss and   (*b*) Eddy current loss

2. Copper losses

    (*a*) Armature winding   (*b*) Field winding

    (*c*) Brush contact and resistance loss (*d*) Interpole field

    (*e*) Compensating winding

3. Mechanical Losses

Frictional and windage losses

4. Stray losses

    (*a*) Commutational   (*b*) Field distortion

The iron losses occur due to alternating flux linkages with the armature and are dependent on magnitude of flux and its frequency.

The mechanical losses are due to rotation and include bearing and brush friction and windage losses. These losses can be determined as the power consumed by unexcited machine at rated speed on no load. This loss is a function of speed. The brush friction loss depends upon the brush material, brush pressure and speed.

The iron and mechanical losses are considered as constant losses and determined by no-load test.

The copper losses are current dependent and also vary with the temperature of windings. The winding resistance is temperature dependent. The hot resistance of windings are normally higher than cold resistance. This variation depends upon the temperature coefficient of windings.

The stray losses depend upon the current in the short circuited coil undergoing commutation and the flux distortion due to armature reaction. Due to armature reaction additional losses takes place in armature teeth. It is difficult to measure these losses. Normally, it is considered 1% of machine output. The stray losses are neglected for small machines.

Circuit Diagram

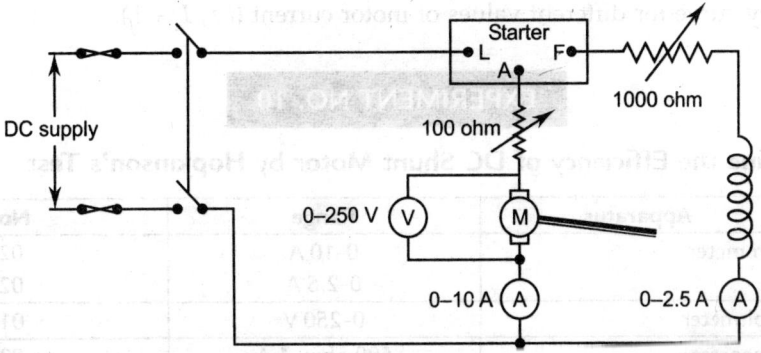

FIGURE 2.74. *Connection diagram for Swinburn's Test*

Procedure

1. Measure armature resistance (R_a) and field circuit resistances (R_f).
2. Perform no-load test (ref. Figure 2.74)

   (*a*) Before applying the dc voltage to machine ensure that there is no external resistance in armature or field circuits. Also there is no load on the machine shaft.

   (*b*) Start the motor with starter.

   (*c*) Adjust the external resistance connected in series with armature such that the voltage across the armature is V.

   V is the difference between rated terminal voltage and full load armature resistance and brush drops.

   (*d*) Adjust the external resistance connected in series with field to obtain rated speed.

   (*e*) Check armature voltage V and re-adjust if necessary.

   (*f*) Note down armature current.

   (*g*) Repeat steps (*c*) to (*f*) for various values of voltage across the armature.

   (*h*) Stop the motor.

Observation Table

S. No.	V	I_a	I_f	$I_m = I_a + I_f$	$I_a^2 R_a$ (1)	$I_f^2 R_f$ (2)	W_o (3)	Total Losses (1)+(2)+(3)	Efficiency η

Calculations

1. Let R_a is the total armature circuit resistance under hot conditions and it includes the armature, compensating winding and interpole winding resistance.
2. The total losses $W = W_o + I_a^2 R_a + I_f^2 R_f + I_a x$ (brush contact drop).
3. The efficiency $\eta = 1 - (W/V . (I_a + I_f))$.

Results

Plot the efficiency curve for different values of motor current (*i.e.*, $I_a + I_f$).

EXPERIMENT NO. 10

2.4.8 Determine the Efficiency of DC Shunt Motor by Hopkinson's Test

S. No.	Apparatus	Range	No.
1.	Ammeter	0–10 A	02
		0–2.5 A	02
2.	Voltmeter	0–250 V	01
3.	Rheostat	500 ohm, 1 A	02
4.	Switch	SPST Type	01

Theory

This test essentially requires two dc machines coupled electrically as well as mechanically. One of the machines runs as motor and give mechanical power to the other machine, which operates as generator. The generator is also electrically coupled with motor. Hence, the power taken from the supply is to overcome the losses incurred in both the machines. The loading on the two machines may be adjusted to any desirable value by changing the field current of these machines.

This method is especially suited for large size machines. Another advantage is that the temperature rise may be determined without much wastage of power. But the main drawback lies in the requirement of two identical machines.

Procedure

1. Connect two mechanically coupled dc machines as shown in Figure 2.75.
2. Before application of DC supply ensure that switch S_1 is open and both external field resistances are short circuited.
3. Switch on the supply and start the motor.
4. Run the motor at rated speed.
5. Check the voltmeter reading of V_1 and if it is nearly double the supply voltage then the polarity of generator is wrong and reverse the polarity. If voltmeter reading is nearly zero, then polarity is correct.
6. Adjust the terminal voltage of generator such that voltmeter reading is zero, then close the switch S_1.
7. Now slightly increase the generator current by reducing the generator field resistance slightly.
8. Now record the ammeter and voltmeter readings.
9. Repeat step 7 and 8 for different values of I_1 and I_2.

Observation Table

S. No.	I_1	I_{1f}	I_2	I_{2f}	I	V

Connection Diagram

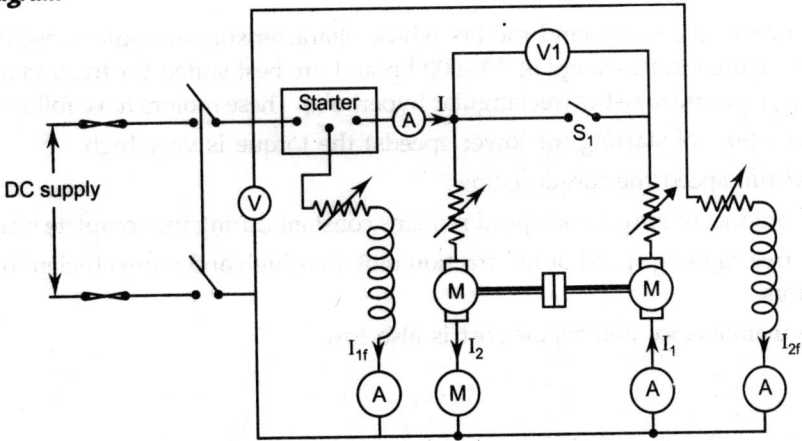

FIGURE 2.75. Connection diagram for Hopkinson's Test

Calculations

Let R_a – is the hot resistance of armature of either machine.

I_1 and I_2 are the armature currents of these machines.

$I_1^2 R_a$ and $I_2^2 R_a$ are the armature copper losses in generator and motor.

VI - is the power drawn from supply.

The total rotational and iron losses $W_o = VI - R_a (I_1^2 + I_2^2)$.

The generator output $P_g = VI_1$

Total generator losses $W_g = I_{1f}^2 R_f + I_1^2 R_a + W_o/2$

Generator efficiency $\eta_g = \left(1 - \dfrac{Wg}{Pg + Wg}\right)$

Motor Input $P_m = V(I_2 + I_{2f})$

Total motor losses $W_m = I_{2f}^2 R_f + I_2^2 R_a + W_o/2$

Motor efficiency $\eta_m = \left(1 - \dfrac{Wm}{Pm}\right)$.

S. No.	VI	$I_1^2 R_a$	$I_2^2 R_a$	W_o	$I_{1f}^2 R_f$	W_g	$P_g = VI_1$	η_g	P_m	W_m	η_m

Results

Plot the efficiency curve for different loading conditions for motor and generator.

EXPERIMENT NO. 11

2.4.9 To Study the Drum Controller and Draw the Speed–torque Characteristics of Different Notches of the D.C. Traction Motor

Necessity of the Experiment

The traction motors are compound motors whose characteristics are quite close to the d.c. series motors manufactured in the range of 300-600 hp and are best suited for traction applications due their self releasing characteristics (rectangular hyperbola). These motors have following advantages:

(i)   At the time of starting (or lower speeds) the torque is very high.

(ii) Near full speed the torque is low.

(iii) The product of torque and speed remains constant during the complete range of operation.

(iv) The free running speed of DC traction motors is high and more efficient than AC traction motors.

(v)   The maintenance and repair cost is also low.

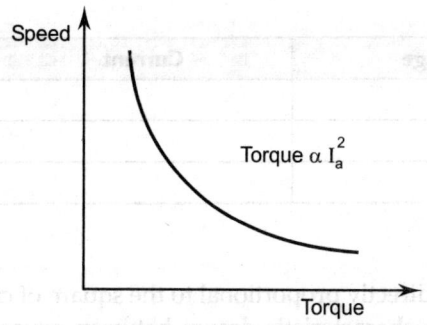

FIGURE 2.76. *Speed torque characteristic of DC series motor*

Drum Controller

It is used for starting, speed control and braking of electrical traction motor in electric trains. The various parts of drum controller are Fingers, Brass segments, and Blow out coil. The construction of fingers resembles human fingers and performs dual function of:

(*i*) Giving uniform and appropriate pressure on segments.

(*ii*) Protecting the brass segments from oxidation.

The function of blow out coil is to extinguish the arc formed due to braking of electrical circuit due to movement of drum. The cross section of the resistance of drum controller is such that, it can carry full load current continuously without excessive heating.

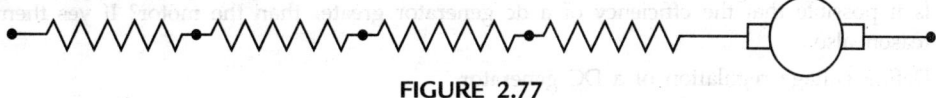

FIGURE 2.77

Connection Diagram

The traction motor has eight terminals for the following four windings (ref. Figure 2.78) :

(*i*) Armature winding   A and AA
(*ii*) Inter pole winding   H and HH
(*iii*) Shunt filed winding   Z and ZZ
(*iv*) Series field winding   Y and YY

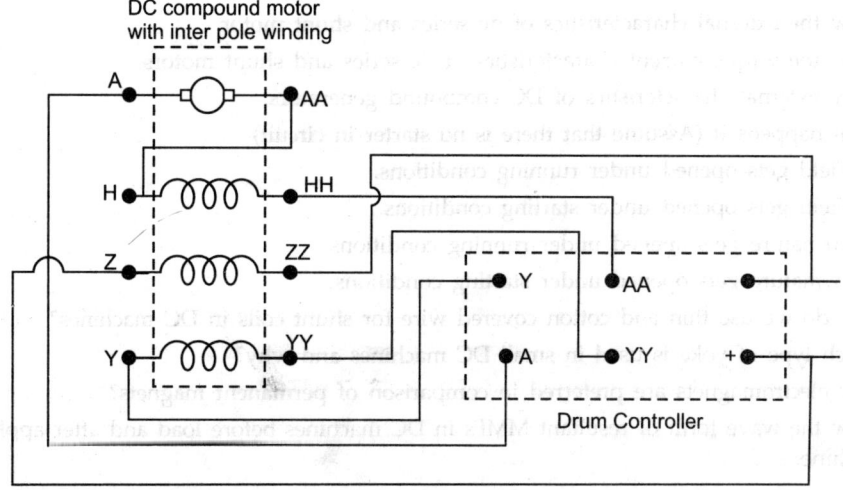

FIGURE 2.78. *Connection diagram for DC compound machine (Traction motor)*

Observation Table

S. No.	Voltage	Current	Speed
1.			
2.			
3.			

Results

In DC series motors, torque is directly proportional to the square of current. Hence, the speed-torque characteristic is similar to the characteristic drawn between square of current and speed.

Precaution

1. Drum Controller handle should be moved from one notch to the other notches, only when the speed has become stable at previous notch.
2. There should always be some water in the pulley to keep it cool.

2.4.10   Review Exercises

I. **Short Answer Questions :**

1. What is the relationship between the mechanical degrees and electrical degrees?
2. What will be the pole pitch of a two-layer lap wound 4-pole dc machine with 16 coils?
3. Is it possible that the efficiency of a dc generator greater than the motor? If yes then mention reason also.
4. Define voltage regulation of a DC generator.
5. In series parallel speed control of a dc series motors the torque in series arrangement is same/half/one fourth/four times the torque in parallel arrangement. (Tick the correct answer).
6. Name the protections, which are provided in a three-point starter of dc shunt motor.
7. Name the protections, which are provided in a starter of dc series motor.
8. Write the following conditions for dc shunt motor
   (a) Maximum efficiency
   (b) Maximum power developed.
9. Draw the external characteristics of dc series and shunt motor.
10. Draw the torque current characteristics of dc series and shunt motors.
11. Draw external characteristics of DC compound generators.
12. What happens if (Assume that there is no starter in circuit)
    (a) Field gets opened under running conditions.
    (b) Field gets opened under starting conditions.
    (c) Armature gets opened under running conditions.
    (d) Armature gets opened under starting conditions.
13. Why do we use thin and cotton covered wire for shunt coils in DC machines?
14. Which type of yoke is used in small DC machines and why?
15. Why electromagnets are preferred in comparison of permanent magnets?
16. Draw the wave form of resultant MMFs in DC machines before load and after applying load on machine.

17. Mention the effect of speed on magnetizing characteristics of DC machine. Also draw the characteristics for different speeds.

18. Differentiate between :

    (a) Rotating and alternating magnetic fields.

    (b) GNA and MNA.

    (c) Lap and wave winding

    (d) Full pitched and chorded winding

    (e) Integral slot and fractional slot

19. Mention three means by which commutation can be improved.

20. Define cross magnetizing and demagnetizing armature reactions.

21. What is fundamental difference between dc motors and synchronous motors?

22. Write two main factors to choose the wave and lap windings.

23. What are the factors which determine the number of brushes and number of parallel paths in wave windings?

24. Define :

    (a) Electrical Efficiency   (b) Mechanical Efficiency

    (c) Commercial efficiency.

25. Name the generators which have

    (a) Drooping characteristic   (b) Rising characteristics

26. Define :

    (a) Continuous rating   (b) Intermittent rating.

27. Why does the speed of DC shunt motor fall as it is loaded?

28. What do you mean by level compounding in DC machines?

29. Name the method which is called as constant torque and variable power method of speed control.

30. What are the purposes of end frames in DC machines?

31. Name the method which is used for speed control higher than its base speed in DC shunt motors.

32. Why the elements of a DC shunt motor starter is graded?

33. Is the revolving magnetic field produced when a two phase supply is applied to the two phase windings displaced by 180 degrees in space.

II. Multiple Choice Questions :

1. Why is the pole shoe in a dc machine larger than its pole body

    (a) It gives sinusoidal flux density

    (b) It provides a support for field winding

    (c) It reduces iron loss in the pole shoes and gives a more nearly rectangular flux density wave

    (d) It helps to make the flux density wave nearly sinusoidal.

2. The armature reaction MMF in a dc machine :

    (a) are in the same direction as the main poles

    (b) are in the direct opposition to the main poles

    (c) make an angle of 90 deg. with the main pole axis

    (d) make an angle with main pole axis which is load dependent.

3. A dc series motor has linear magnetisation and negligible armature resistance. The motor speed is :

    (a) directly proportional to square root of T

    (b) inversely proportional to square root of T

(c) directly proportional to T

(d) inversely proportional to T

4. The armature reaction MMF in a dc machine is :

(a) sinusoidal in shape

(b) trapezoidal in shape

(c) rectangular in shape

(d) triangular in shape.

5. The armature reaction mmf wave in a dc machine :

(a) is stationary relative to the brushes

(b) moves relative to the brushes at armature speed

(c) moves relative to the brushes at synchronous speed.

6. The process of current commutation in a dc machine is opposed by the

(a) emf induced in the commutation coil because of the inter-pole flux.

(b) reactance emf

(c) coil resistance

(d) brush resistance

7. Field control of dc shunt motor gives

(a) constant torque drive

(b) constant power drive

(c) constant speed drive

(d) variable load speed drive.

8. A sinusoidal voltage of frequency 1 Hz is applied to the field of dc generator. The armature voltage will be :

(a) 1 Hz square wave

(b) 1 Hz sinusoidal wave

(c) DC voltage

(d) none of the above.

9. For equal power rating generators, the total current is more in which generator :

(a) wave wound

(b) lap wound

(c) simplex wound

(d) none of the above.

10. If a DC motor is connected across AC supply, the DC motor will

(a) burn as the eddy currents in the field produce heat.

(b) run at its normal speed.

(c) run at lower speed.

(d) run continuously but sparking takes place at brushes.

11. What would be the standard direction of rotation of induction motor when looking at the front of the motor

(a) clockwise

(b) anticlockwise

(c) in any direction

(d) none of the above.

12. The efficiency of DC generator is maximum when

(a) magnetic losses equal to mechanical losses

(b) armature copper losses equal to constant losses

(c) field losses equal to constant losses

(d) stray losses equal to copper losses.

13. Equalizer rings in DC generator (lap wound) are used

(a) to avoid unequal distribution of current at the brushes thereby helping to get sparkles commutation.

(b) to avoid harmonics developed in machine

(c) avoid noise

(d) all above.

14. The sparking at the brushes in DC generator is attributed to
    (a) quick reversal of current in the coil under commutation
    (b) armature reaction
    (c) reactance voltage
    (d) high resistance brushes.

15. In series parallel control method when two DC series motors are connected in series, the speed of the set is
    (a) half of the speed of the motors when connected in parallel.
    (b) one fourth of the speed of the motors when connected in parallel.
    (c) same as in parallel
    (d) rated speed of any one motor.

16. The torque produced by series combination of two series motors is
    (a) equal to the torque when they are connected in parallel.
    (b) half of the torque when they are connected in parallel.
    (c) four times the torque when they are connected in parallel.
    (d) two times the torque when they are connected in parallel.

17. The speed of DC series motor increases as the armature torque increases
    (a) True   (b) False

18. When the torque of DC series motor is doubled the power is increased by
    (a) 70 %   (b) 50 to 60 %   (c) 20 %   (d) 100 %.

19. Swinburn's test is applied to
    (a) those machines in which flux is practically constant.
    (b) those machines in which flux is varying
    (c) those machines in which flux is proportional to armature current
    (d) none of the above.

20. The retardation test is applicable to shunt motors and generators and is used to find
    (a) the stray losses   (b) the copper losses
    (c) the eddy current losses   (d) the frictional losses.

21. If the iron losses and full load copper losses are given then the load at which two losses are equal is given by
    (a) FL × iron losses/copper losses   (b) FL × (iron loss)^2/FL copper loss
    (c) FL × √(Iron loss/Cu loss)   (d) FL × √(Cu loss/Iron loss).

22. What will happen to the DC generator if the field winding attains the critical resistance
    (a) It will generate maximum voltage   (b) it will generate maximum power
    (c) it will not develop voltage at all   (d) non of the above.

23. The thickness of laminations used for armature core of DC machine is of the order of
    (a) 0.005 mm   (b) 0.05 mm   (c) 0.5 mm   (d) 5 mm.

24. The conductors of DC machine armature are
    (a) welded   (b) soldered to armature
    (c) firmly placed in slots   (d) Wound round the armature core.

25. A two layer lap wound 4 pole generator with 16 coils will have a pole pitch of
    (a) 32   (b) 16   (c) 8   (d) 4.

26. The residual magnetism of a DC shunt generator can be regained by
    (a) Connecting the shunt field to a battery
    (b) Running a generator on no load
    (c) Earthing the shunt field
    (d) Reversing the direction of the generator.

27. Out of the following losses in a DC generator which one has the least proportion
    (a) Copper loss   (b) Hysterisis losses
    (c) Eddy Current losses   (d) Windage losses.

28. Pole shoes of DC machines are fastened to the pole core by
    (a) Welding   (b) Soldering
    (c) Wooden buttons   (d) Counter sunk screw.

29. Each Commutator segment is connected to armature conductor by means of
    (a) Insulator   (b) Dielectric   (c) Copper lug   (d) Resistance wire.

30. Which of the losses will increase rapidly as compared to the others, when speed of machine is increased
    (a) Friction loss   (b) Copper losses   (c) Hysterisis loss   (d) Eddy current loss.

31. The direction of rotation of DC motor can be reversed by
    (a) Reversing the field current   (b) Reversing the armature current
    (c) Change the polarity of DC supply   (d) Both (a) and (b).

32. An amplidyne can be used as
    (a) DC series motor   (b) DC shunt motor
    (c) DC compound motor   (d) Magnetic amplifier

33. For equal output machines; the total current is more in
    (a) Wave wound machine   (b) Lap wound machine
    (c) Simplex winding machine   (d) None of these.

34. It is preferable to start a dc series motor with some mechanical load on it because
    (a) It may otherwise develop excessive speed and demage itself
    (b) It will not run at no load
    (c) A little load will act as a starter to the motor
    (d) None of the motor.

35. The function of the commutators in DC machine is
    (a) To change alternating current to direct current
    (b) To improve commutation
    (c) For easy speed control
    (d) To change alternating voltage to direct voltage.

36. The overload coil in a DC motor starter is connected in series with .... and made by .... wire.
    (a) Shunt, thin   (b) Shunt, thick   (c) Armature, thin   (d) Armature, thick.

III. Fill in the Blanks :

1. If the brushes of a DC machine are shifted in the forward direction, if the machine is operating in generating mode then the flux per pole will .... and when it will operate in motoring mode the flux per pole will ....

2. Name the materials which are used for the following parts of a large DC machine
   (a) Commutator segments ....
   (b) Insulation between commutator segments ....
   (c) Brushes ....
   (d) Armature core ....
   (e) Yoke ....

3. The main field wave is .... in shape and armature flux wave is .... in shape.

4. The equalizer rings required in .... wound generators.

5. The lap wound machines is .... voltage and .... current machines.

6. The wave wound machines is .... voltage and .... current machines.

7. The number of mechanical and electrical degree for a dc generator will be the same when the generator has .... poles.

8. If the shunt field is connected across armature the compound generator is said to have .... connections.

9. If the back pitch is less than front pitch the winding is said to be ....

10. Both front pitch and back pitches in a dc armature are always ....

11. If there is no residual magnetism a .... excited generator cannot build up its emf.

12. In a three-point starter for dc motors the no-volt coil is connected in .... with shunt field coil and made by .... wire.

13. In a four point starter for dc motors the no volt coil is connected in .... with ....

14. The value of last element of dc shunt motor starter is .... the first element.

15. In wave winding the number of brushes is equal to ....

16. In lap winding the number of brushes is equal to ....

17. In DC Machines the brushes remains in contact with conductors under .... (north/south/inter poles).

18. The voltage developed is maximum at the center of north and south pole/inter-pole/none.

19. DC series motor with starter, if the field circuit gets opened under running conditions the motor speed will be ....

20. DC series motor without starter, if the field circuit gets opened under running conditions the motor speed will be ....

Answers

II. Multiple Choice Questions :

1. (c)	2. (c)	3. (b)	4. (d)	5. (a)
6. (b)	7. (b)	8. (b)	9. (b)	10. (a)
11. (b)	12. (b)	13. (a)	14. (c)	15. (b)
16. (c)	17. (b)	18. (a)	19. (a)	20. (a)
21. (c)	22. (c)	23. (c)	24. (c)	25. (c)
26. (a)	27. (d)	28. (d)	29. (c)	30. (d)
31. (d)	32. (c)	33. (b)	34. (a)	35. (d)
36. (d)				

III. Fill in the Blanks :

2. (a) Hard drawn copper   (b) Mica   (c) Graphite or soft copper   (d) Si-steel   (e) Cast iron

3. Trapezoidal, triangular   5. Low, high   6. high, low

7. 2   8. Short shunt compounded generator

10. odd number   11. self   12. series, thin   13. series, resistance

14. lesser   15. Two   16. No. of poles

18. North and south pole   19. zero   20. infinite

SECTION–C

2.5 SYNCHRONOUS MACHINE

2.5.1 Introduction

There are three significant families of electrical rotating machines.

(a) Synchronous machines, (b) Asynchronous machines, and (c) D.C. Machines

The three phase synchronous generators are the primary source to all the electrical energy supplying systems. The size of synchronous generators varies from a few watts of domestic generators to thousands of MW in generating stations. The armature winding of large synchronous machine is normally on the stator and field winding on rotor due to economic reasons. The field winding is a low power winding and hence, it is easy to supply low power. The armature and field laminations are shown in Figure 2.79.

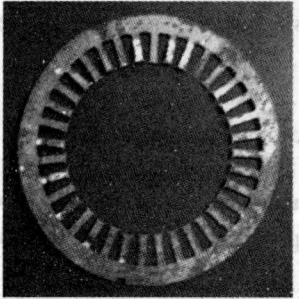

(a) Armature core Silicon steel Lamination for stator

(b) Lamination for Field core of rotor

(c) Synchronous Machine consisting of laminated armature

FIGURE 2.79. *Stampings for distributed stator winding and Projected Field rotor poles.*

2.5.2   Classifications of Synchronous Machine

Synchronous machines may be classified as :

  A. According to their excitation system :

  There are two types of synchronous generators:

   (a) Brush and slip ring type alternator

   (b) Brushless alternator.

The d.c. power is to be supplied to field winding, which is on the rotor through brush and slip ring arrangement. In this arrangement continuous wear and tear takes place, which requires continuous maintenance. Hence, brushless type rotors are devised.

  B. According to the rotor construction :

   (i) Salient or projected pole rotor

   (ii) Cylindrical or round rotor

The salient pole machine is normally used for low speed prime movers (hydro-turbines) and called hydro-alternators. In these type of alternators the rotor diameter is larger than its length (D/L ratio is high).

On the other hand, cylindrical rotor machine is normally high speed machines. It is driven by high speed prime movers like steam turbine. Due to the high speed the rotor diameter is smaller, otherwise, due to centrifugal force the windings will come out from the rotor slots. Hence, in these type of alternators the rotor diameter is smaller and length is large (i.e., D/L ratio is low). This type of alternator is called turbo alternators.

2.5.3   Principle of Operation of Synchronous Motor

In synchronous motor d.c. power is supplied to the rotor and a.c. power to the stator, however, in induction motor the a.c. power is supplied to stator and power transferred to rotor is through induction. When the d.c. power is supplied to rotor of synchronous motor, it creates a rotor flux. This rotor flux interacts with the rotating magnetic field of the stator which in turn causes the rotor to rotate. A steady state speed of the rotor is the same as the speed of the rotating magnetic field in the stator i.e., synchronous speed, n_s, hence, it is called synchronous motors.

2.5.4   Characteristics of Synchronous Machine

The influence of armature reaction upon the variation of synchronous machine terminal voltage with load for constant field current is shown in Figure 2.80. When the power factor of the load is unity, the fall in voltage with increase in load is comparatively small as compared to inductive load. With an inductive load, the demagnetizing effect of armature reaction of stator is in opposite direction as that of rotor flux, which decreases the net airgap flux, causes the terminal voltage to fall much more rapidly. With a capacitive load, the magnetizing effect of armature reaction of stator is in the same direction as that of rotor mmf and increase the net airgap flux, which causes the terminal voltage to increase with increase in load.

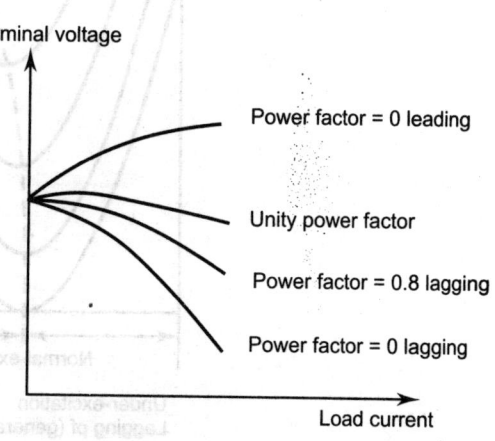

FIGURE 2.80. *Variation of terminal voltage with load*

2.5.5 Effect of Excitation at Constant Load

It is observed that at constant load as excitation emf E_f is varied (by varying filed current I_f), the power angle δ varies such that $E_f \sin \delta$ remains constant as shown in Figure 2.81. The excitation corresponding to unity power factor is known as normal excitation, while excitation larger than this is known as overexcitation and less than this is called under excitation.

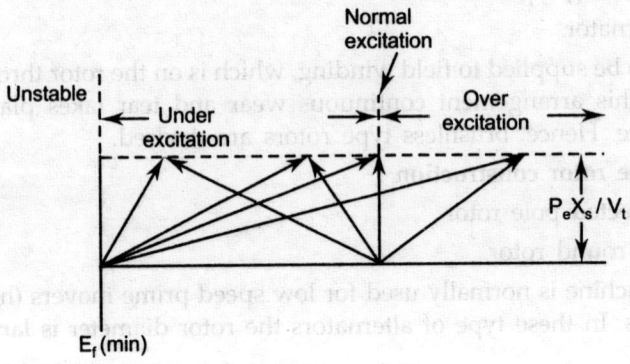

FIGURE 2.81. *Effect of excitation at constant load*

2.5.6 V-curves for Synchronous Machine

In synchronous machine the real power exchanged with bus bars is controlled by the mechanical shaft power irrespective of excitation. The excitation, on the other hand governs only the power factor of machine by controlling the reactive power exhange with the system as shown in Figure 2.82. If it is desired to feed more power to the bus bars the input to the turbine is increased, which in turn increase mechanical input to the shaft. This will increase the power angle δ and so does the power output.

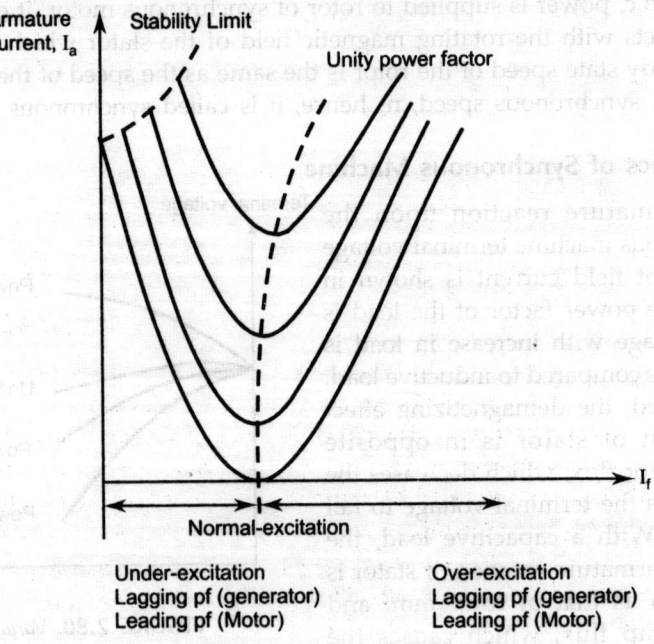

FIGURE 2.82. *V-curves of synchronous machine (Constant load and variable excitation)*

Power-angle Characteristic of Synchronous Machine

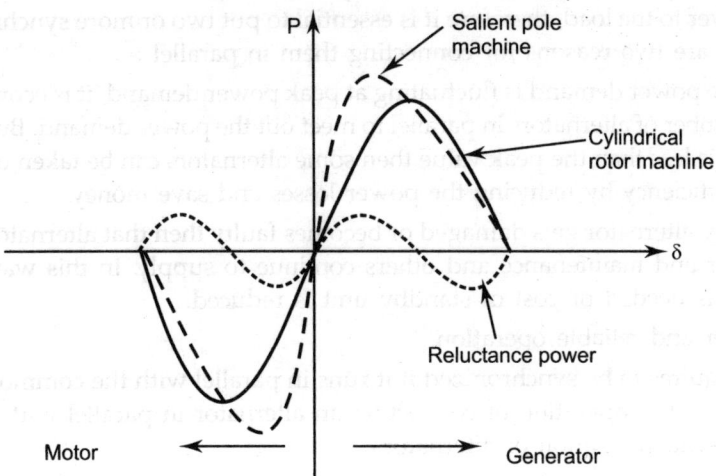

FIGURE 2.83. *Power-angle characteristic of synchronous machines*

Short Circuit Transients in Synchronous Machine

The short circuit transients shown in Figure 2.84 for synchronous machine is a highly complex phenomenon because there are number of circuits coupled together.

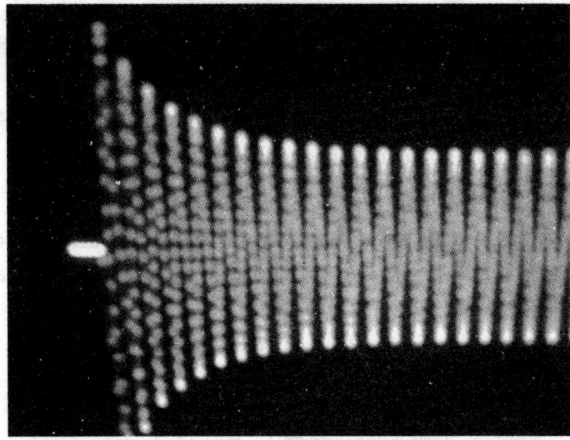

FIGURE 2.84. *Current pattern for 3-phase dead short circuit on generator*

2.6   EXPERIMENTS ON SYNCHRONOUS MACHINE

EXPERIMENT NO. 12

2.6.1   To Study the Methods of Synchronization and Also Synchronization of the Alternator with Bus Bar by anyone Method

Necessity of the Experiment

Now-a-days the power demand is intensively increasing and one alternator is not sufficient to deliver the power to the load, therefore it is essential to put two or more synchronous generators in parallel. There are two reasons for connecting them in parallel :

(*i*) When the power demand is fluctuating at peak power demand, it is economical to connect more number of alternators in parallel to meet out the power demand. But when the power demand is less than the peak value then some alternators can be taken out to improve the system efficiency by reducing the power losses and save money.

(*ii*) When any alternator gets damaged or becomes faulty then that alternator can be removed for repair and maintenance and others continue to supply. In this way the less reserve capacity is needed or cost of standby unit is reduced.

(*iii*) For better and reliable operation.

A machine requires to be synchronized if it runs in parallel with the common bus bar or with the other alternator. The operation of connecting an alternator in parallel with the common bus bar is known as synchronization of alternators.

Theory

Before synchronization, following conditions must be satisfied:

1. **Equality of Voltage :** The terminal voltage of both the systems *i.e.*, the incoming alternator and the bus bar voltage or other alternator must be same.

2. **Synchronism of Phase :** The phase sequence of both the systems must be same.

3. **Equality of Frequency :** The frequency of both the systems must be same. The condition (1) can be checked with the help of voltmeter and the condition (2) and (3) by any synchronizing method. There are three synchronizing methods :

(*a*) Using incandescent lamp, (*b*) Using synchroscope, (*c*) Using Automatic synchronizer.

Now we discuss in detail about these methods.

(*d*) Using Incandescent lamp

Let machine G_2 be synchronized with machine G_1 which is already connected with the bus bar, using three lamps (L_1, L_2 and L_3) method. These lamps are known as synchronizing lamps connected as shown in Figure 2.85.

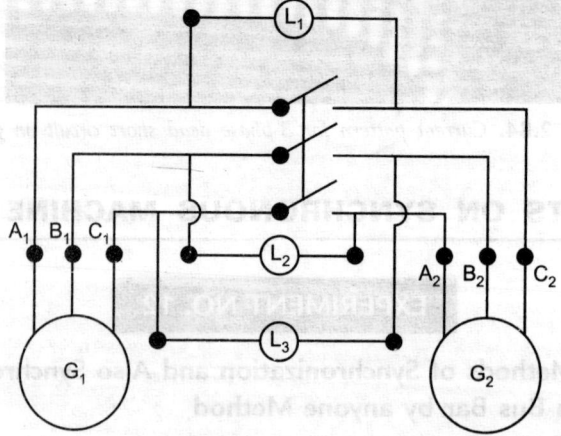

FIGURE 2.85. *Synchronization using three lamp method*

If the speed of machine 2 is not brought upto that of machine 1 then its frequency will also be different, hence there will be a phase difference between their voltages as shown in Figure 2.86. Due to difference in frequencies the resultant voltage $(e_1 + e_2)$ will under go changes similar to the frequency changes of beats produced when two sound sources of nearly equal frequencies are sounded together. The resultant voltage is sometime maximum and sometimes minimum. Hence the lamps will flicker, sometime dark and sometimes bright. Synchronization is done at the middle of the dark period. This method of synchronizing is known as dark lamp method.

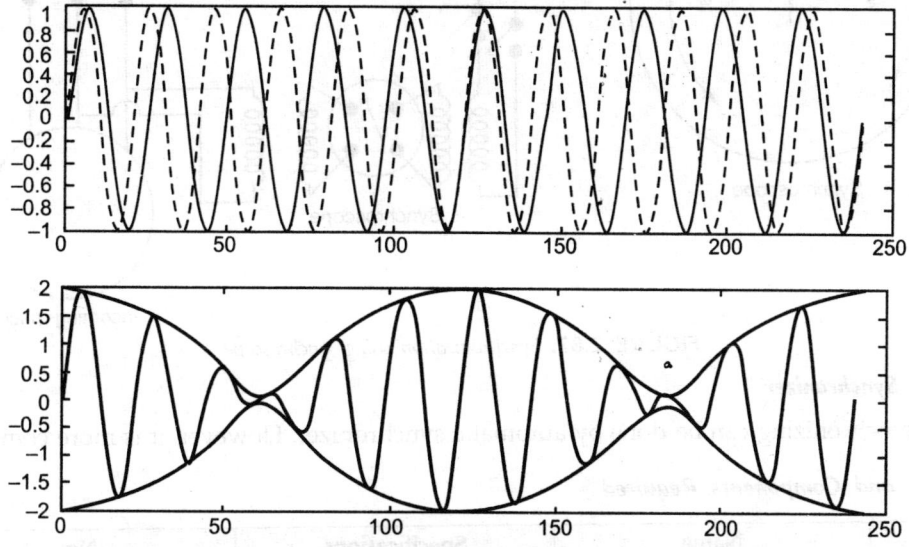

FIGURE 2.86. *Waveforms when two systems operating at different frequencies.*

Lamp L_1 is connected between A_1 and A_2, L_2 between B_1 and C_2 and L_3 between C_1 and B_2. These three lamps slowly brighten and darken in cyclic successor in a direction depending upon whether incoming machine 2 is fast or slow. The synchronizing switch will be closed at the moment when lamp L_1 will be completely dark. This transposition of two lamps suggested by Siemens and Aalske helps to indicate whether the incoming machine 2 is running too slow or too fast. If lamps were connected symmetrically, they would dark out or glow up simultaneously (if phase rotation is same.).

It should be noted that synchronizing by dark lamp method has following drawbacks.

1. The lamps become dark at about one third of the rated voltage. Hence, faulty synchronizing may be done in dark period.

2. Using lamp method it can not find out that how much the machine is slow or fast.

3. This method is not applicable for high voltage alternators, because lamp ratings are normally low. For such situations we need an extra transformer to step down the voltage.

Synchronizing by Synchroscope

Synchroscope is a device that shows the correct instant of closing the synchronizing switch with the help of a pointer which will rotate on the dial. The rotation of pointer also indicates whether the incoming machine is running too slow or too fast. If incoming machine is slow then pointer rotates in anticlockwise direction and if machine is fast then pointer rotates in clockwise direction as shown in Figure 2.87.

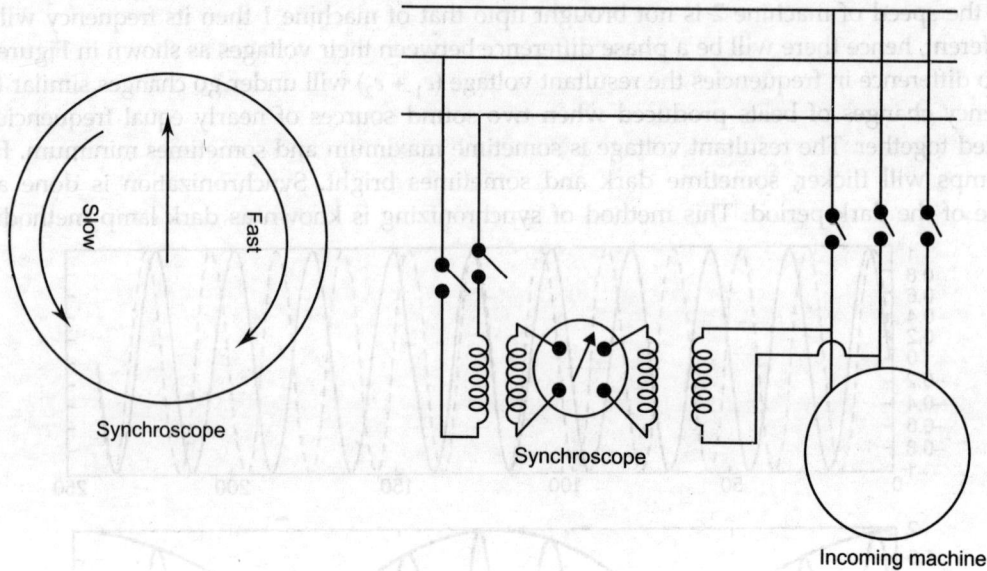

FIGURE 2.87. *Synchronization using synchroscope*

Automatic Synchronizer

The best synchronizing can be done by automatic synchronizer. However, it is more complicated.

Equipment and Components Required

S. No.	Name	Specifications	No.
1.	Alternator	Rated voltage =	1
		Rated current =	
		Rated hp	
		No. of pole =	
		Rated frequency =	
		No. of phases =	
2.	Voltmeter		1
3.	Rehostate		2
4.	Techno meter		1

Connection Diagram

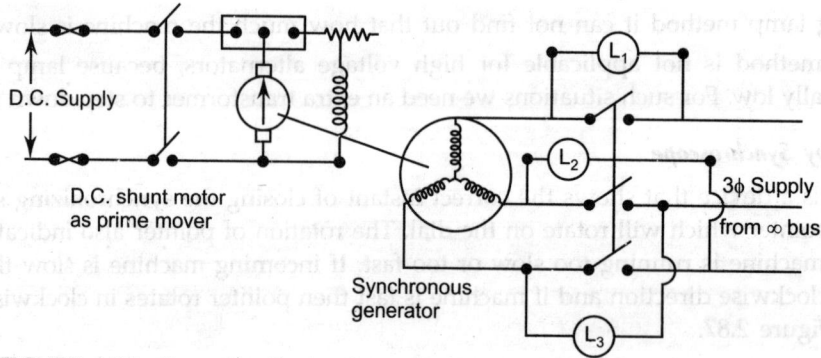

FIGURE 2.88. *Connection diagram for synchronizing the alternator with 3-lamp method*

Procedural Steps

1. Make the connection diagram as shown in Figure 2.88.
2. Run one of the alternators and adjust its voltage at rated value and close switch to bus bar.
3. Start the second set (incoming machine 2), bring it upto proper speed equal to that of the running machine (bus bar voltage).
4. Synchronize the incoming machine by any one method described above.

2.6.2 Experimental Quiz

1. What are essential conditions for synchronization of alternator with the infinite bus bar?
2. Why it is necessary to synchronize the alternator?
3. What are the advantages of parallel operation of the alternators?
4. Name the methods of synchronization.
5. Which method would you like to prefer and why?
6. In dark lamp method, why do we do the transposition of lamps?
7. Can we do synchronization without transposition? if yes then how?
8. What are the advantages and disadvantages of the dark lamp method?
9. A synchronous machine is synchronized with an infinite bus and then by changing its field current it is made to operate as (a) a motor, (b) as a generator. Whether the machine is delivering or absorbing reactive power?
10. What is the effect of wrong synchronization?
11. In synchronization the incoming machine experience a jerk on account of wrong synchronizing with bus bar. Why?
12. An alternator has a phase sequence of ABC for its three phase output voltages. What will be the phase sequence if field current is reversed?
13. An alternator has phase sequence ABC for its three output voltages, for clockwise rotation. What will be the phase sequence if it rotates in the anticlockwise direction?

EXPERIMENT NO. 13

2.6.3 To Perform (1) No Load Test and (2) Short Circuit Test on the Alternator and Determine Regulation by E.M.F. Method and M.M.F. Method

Necessity of the Experiment

Now a days, A.C. power is supplied to the consumers and D.C. power system is totally vanished. The reason of the above change is that A.C. voltage level can be increased to a very high value like 750 kVs, 1000 kVs, 1500 kVs and so on and when voltage level increases, the transmission losses are reduced tremendously. For generating A.C. power we use large synchronous generators in generating stations. When load is applied on the alternator (synchronous generators) the terminal voltage changes due to impedance drop of machine. But the consumers voltage must be maintained within prespecified limits of voltage (±10 % of rated voltage). This demands that the machine is to be designed with low voltage regulation. But a machine with low voltage regulation is uneconomical and is subjected to large mechanical and electrical stresses in case of short circuits. Hence, the voltage regulation is an important characteristic of an alternator and its predetermination is essential for its normal operations as well as for designing suitable excitation control scheme.

Theory

Voltage Regulation : The voltage regulation of an alternator is per unit terminal voltage rise when a given load at a given power factor is thrown off, for constant excitation and speed.

$$\% \text{ Voltage regulation } = \frac{(V_{NL} - V_{FL})}{V_{FL}} * 100$$

where

V_{NL} – no load voltage

V_{FL} – full load voltage

NOTE In case of leading power factor, terminal voltage will fall on removing the load. Hence, the voltage regulation is negative, but in lagging power factor loads, terminal voltage rise on removing the load. Hence, the regulation is positive.

Methods of Determination of Voltage Regulation

There are two methods of determination of voltage regulation

(a) Direct method and

(b) Indirect Method.

Direct loading method is only applicable to the small machines. But in case of large machines the application of direct loading is uneconomical and sometimes not possible at all. Therefore, indirect methods are used to determine the voltage regulation of alternators, which are listed below :

1. Synchronous impedance method or E.M.F. method

2. Ampere Turn (AT) method or M.M.F. method.

3. Zero Power Factor (ZPF) method or Potier method

4. New A.S.A. method.

1. Synchronous Impedance Method : This method is based on the simple equivalent circuit and phasor diagram. In this method the effect of armature reaction is expressed as a voltage drop I_a*X_{ar} (X_{ar} is normally called armature reaction reactance).

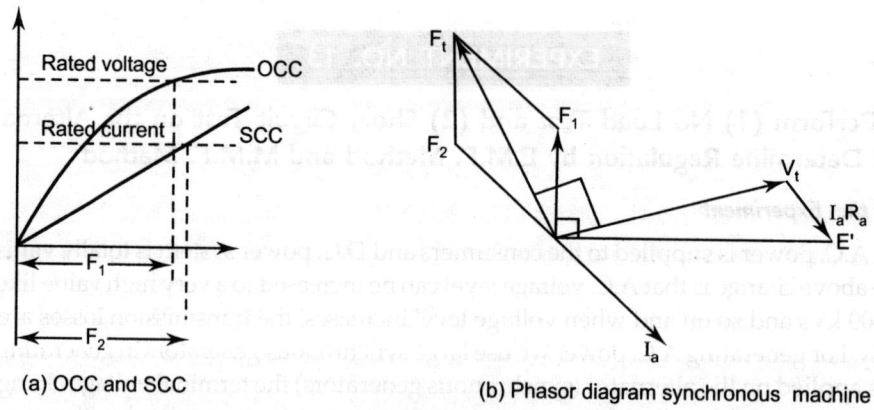

(a) OCC and SCC

(b) Phasor diagram synchronous machine

FIGURE 2.89

F_1 – mmf per pole required to produce rated voltage on open circuit.

F_2 – mmf per pole required to produce rated short circuit current.

The mmf F_1 has to produce the impedance drop in the armature and neutralize the armature reaction mmf and in phase opposition to current I_a.

Equipments and Components Required

S. No.	Name	Specifications	No.
1.	Alternator	Rated voltage =	1
		Rated current =	
		Rated frequency	
		No. of pole =	
		Power KVA =	
		Speed RPM =	
		No. of phases =	
2.	Voltmeter	AC	1
		DC	1
	Voltmeter	AC	1
	Rehostate		1
	Tachometers		1

Connection Diagram

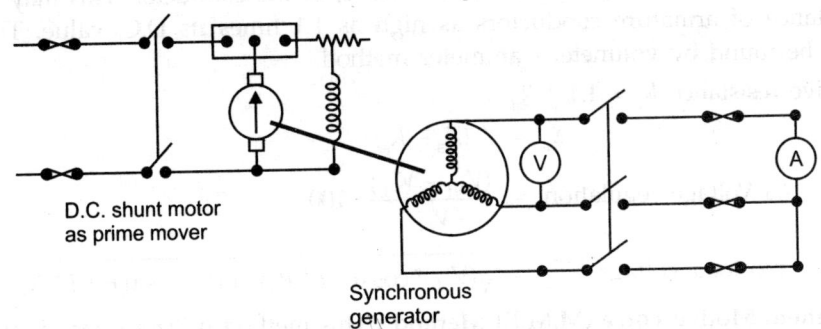

D.C. shunt motor
as prime mover

Synchronous
generator

FIGURE 2.90. *Connection diagram for open circuit and short circuit test on alternator*

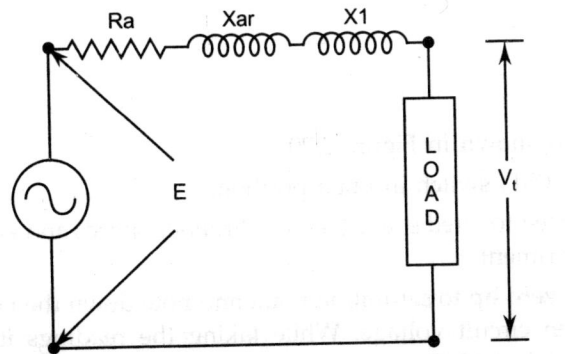

FIGURE 2.91. *(a) Equivalent circuit of Synchronous generator*

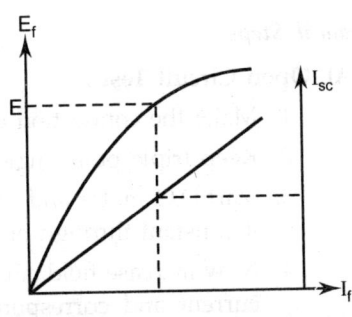

FIGURE 2.91. *(b) OCC and SCC*

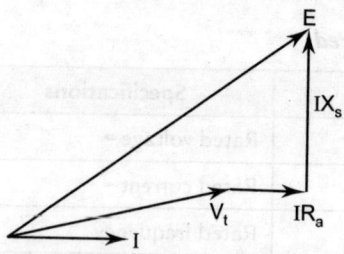

FIGURE 2.91. (c) Phasordiagram

Synchronous reactance $X_s = X_1 + X_{ar}$

Synchronous impedance $Z_s = R_e + j X_s$

where

X_1 – leakage reactance

R_e – effective resistance per phase.

The open circuit voltage corresponding to the field current I_f is E. When winding is short circuited, the terminal voltage is zero. Hence, it may be assumed that the whole of this voltage E is being used to circulate the armature short circuit current I_{sc} against synchronous impedance Z_s.

$$E = I_{sc} * Z_s \rightarrow Z_s = E/I_{sc}$$

As described above the synchronous impedance is the phasor summation of effective resistance and synchronous reactance. The effective resistance is different from the armature resistance, due to skin effect, current tries to flow on the skin (surface) of the conductor. This may increase the effective resistance of armature conductors as high as 1.1 times its D.C. value. The armature resistance can be found by voltmeter - ammeter method.

The effective resistance $R_e = 1.1 * R_{dc}$

$$X_s = \sqrt{(Z_s - R_e)}$$

% Voltage regulation $= \dfrac{(V_{NL} - V_{FL})}{V_{FL}} * 100$

No load voltage $V_{NL} = \sqrt{(V_{FL} * \cos\theta + I * R_a)^2 + (V_{FL} * \sin\theta + I * X_s)^2}$

2. By Magneto Motive Force (M.M.F.) Method : This method utilizes open circuit and short circuit characteristics in determining the regulation. In this method, the effect of X_1 and X_{ar} has been jumped together as an mmf.

Procedural Steps

(A) Open Circuit Test :

1. Make the connection diagram as shown in Figure 2.90.
2. Keep triple pole single throw (TPST) switch in open position.
3. Start D.C. motor and adjust its speed to rated speed (*i.e.*, synchronous speed) and keep it constant through out the experiment.
4. Now increase field current from zero up to saturation point and note down the field current and corresponding open circuit voltage. While taking the readings it is necessary to ensure that the speed of M-G set must be constant.

(B) Short Circuit Test :

1. Connect an ammeter of 20 A range on any two terminals of alternator and short circuit the remaining terminals as shown in Figure 2.90.
2. Start the D.C. motor and run the motor–generator set at rated speed. Keep the speed constant through out the experiment.
3. Gradually increase the field current from zero to rated value.
4. While increasing the filed current note down the reading of field current and armature current (I_a).

(C) Measurement of Armature Resistance :

1. Open all the connections of D.C. motor as well as alternator.
2. Measure the d.c. resistance per phase (R_a) of the stator with the voltmeter-ammeter method.
3. For finding the effective resistance of the armature multiply 1.1 into the D.C. resistance.

Observation

(A) Open Circuit Test :

At constant speed .... rpm.

S. No.	Field current (Amp)	Terminal voltage V_0 (volts)
1.		
2.		
3.		
4.		

(B) Short Circuit Test :

At constant armature current .... Amp

S. No.	Field current (Amp)	Short circuit current *Isc* (Amp)
1.		
2.		
3.		
4.		
5.		

(C) Measurement of Armature Resistance :

At constant armature current .... Amp

S. No.	Voltage (Volts)	Current (Amp)	D.C. resistance $R_a = V/I$
1.			
2.			
3.			
4.			
5.			

Calculation

(A)   1. Plot the short circuit and open circuit characteristic on the same graph paper.

2. Read field current at rated short circuit (I_{sc}) from short circuit characteristics.

3. Read the open circuit voltage (V_o) corresponding to this field current which is obtained in step two from open circuit characteristic.

4. Calculate synchronous impedance.

(B) By M.M.F. Method.

1. Measure F_1 – mmf per pole required to produce rated open circuit voltage from OCC.

2. Measure F_2 – mmf per pole required to produce rated short circuit current from short circuit characteristic

3. Draw phasor diagram as shown in figure

4. Calculate total mmf F.

5. Measure open circuit voltage at F from OCC.

6. Calculate the regulation.

Result

Percentage regulation in E.M.F. method = ...

Percentage regulation in M.M.F. method = ...

From the above experimetnation it can be concluded that the voltage regulation found by E.M.F. method is lesser than the voltage regulation found by M.M.F method, because the M.M.F. method employs Z_s which does not have a constant value. Its value depends on field excitation.

PROBLEM. *Write a Matlab Program for drawing the phasor diagram of synchronous machine (generator mode or motor mode) for given power factor.*

2.6.4   Matlab Program for Drawing Phasor Diagram

```
%******************************************************************
%
% syn_phasor.m - Draws phasor diagram for a round rotor syn. machine
%
%******************************************************************
clear all;
mode=input('Operating mode ( gen or mot ) ','s');
VL=1.0; Ia=1.0; Vt=VL;
Ra=0.05; Xs=0.9; Xl=0.15*Xs;
PF=input('Power Factor ( numerical value ) ');
if PF < 0 | PF > 1; disp('WARNING - Improper PF value'); break;
else; end
if PF ~= 1.0
PFS=input('PF ( leading or lagging ) ','s');
else; PFS='unityPF'; end
```

```
thet=acos(PF);
if PFS == 'lagging'
   Ia=Ia*exp(-j*thet);
elseif PFS == 'leading'
   Ia=Ia*exp(j*thet);
else
   Ia=Ia+j*0;
end
if mode == 'gen'
   p1=0;
   p2=Vt;
   p3=Vt+Ia*Ra;
   p4=Vt+(Ra+j*Xs)*Ia;
   p5=Vt+(Ra+j*Xl)*Ia;
      else
   p1=0;
   p2=Vt;
   p3=Vt-Ra*Ia;
   p4=Vt-(Ra+j*Xs)*Ia;
   p5=Vt-(Ra+j*Xl)*Ia;
end
P=[p1 p2 p3 p4 p1 p5];
R=abs(P);Rm=1.1*max(R);
A=angle(P); figure(1);
polar(0,Rm); hold on; % Size plot scale
polar(A,R); polar(A(2),0.98*R(2),'>'); % Van arrow head
polar(A(4),R(4),'.'); % Dot at tip of Ef
polar(A(6),R(6),'.'); % Dot at tip of Er
if mode == 'gen'; text(0.8*R(2),-0.1*R(2),'Van');
else; text(0.8*R(2),0.1*R(2),'Van'); end
text(1.05*real(p4),1.1*imag(p4),'Ef');
text(1.05*real(p5),1.1*imag(p5),'Er');
if mode == 'gen'
   text(-0.9*Rm,-0.1*Rm,'Generator mode');
else
   text(-0.9*Rm,-0.1*Rm,'Motor mode');
end
text(-0.9*Rm,-0.3*Rm,['PF = ',num2str(PF),' ',PFS]);
hold off;
%********************************************************************
```

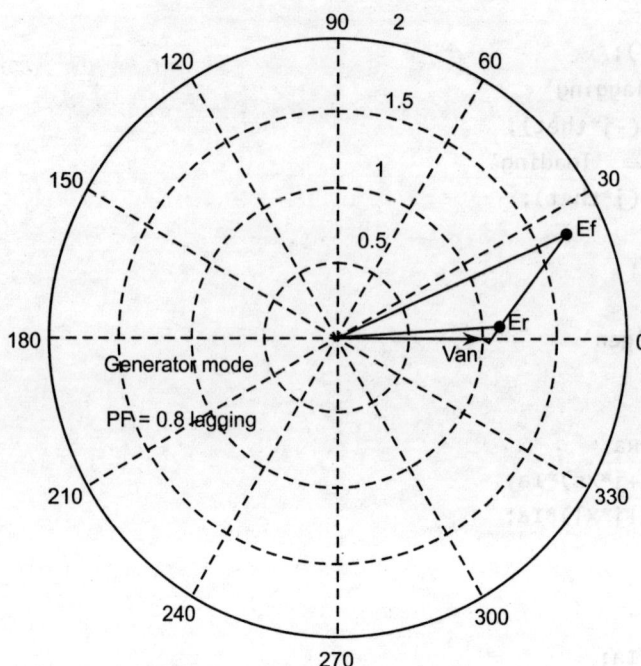

FIGURE 2.92. *Matlab program output*

2.6.5 Experimental Quiz

1. Define regulation of a synchronous generator (alternator).
2. Mention different methods to determine the regulation of alternator in the lab.
3. Which method of finding regulation gives the most acceptable results?
4. Give the reason, why does the synchronous impedance method (E.M.F. method) give regulating higher than actual and potier (M.M.F) method give lesser value?
5. Name the factors which govern the regulation of an alternator.
6. How and why the power factors of load affect the regulation of an alternator?
7. If the field excitation of alternator changes at constant speed then what quantity will change?
8. If the field of D.C. motor changes, then what quantity will change?
9. What are the conditions for performing the open circuit test?
10. What is the power factor of an alternator under short circuit condition?
11. Why is short circuit characteristic a straight line?
12. Up to what range short circuit characteristic is linear?
13. What is the value of voltage at which you perform the short circuit test?
14. How and why is the synchronous reactance affected by the field current?
15. Can we compensate the armature reaction in the synchronous generator as in the D.C. machines?
16. What is the number of poles of the alternator on which you had done the practical?
17. Does the frequency of the alternator depend on the number of poles? If yes then give the relation?
18. How many number of poles are there in high speed turbo alternators and in water wheel generators? Justify your answer.
19. What is the constructional difference between turbo alternator and water wheel generators?
20. Why is regulation an important characteristic of an alternator in the interconnected system?

EXPERIMENT NO. 14

2.6.6 To Perform (1) No Load Test and (2) Zero Power Factor Test on an Alternator and Determine Regulation by Zero Power Factor Method

Necessity of the Experiment

As described in the previous experiment the regulation of an alternator is an important characteristic. So it is necessary to determine percentage voltage regulation accurately. But the values of voltage regulation found by E.M.F. method and M.M.F. method are not exact. In E.M.F. method voltage regulation comes out to be somewhat lesser than its actual value and in the latter method (M.M.F. method) the value is somewhat greater than the actual value. Therefore, zero power factor method is used for getting better result.

Theory

The zero power factor or potier method is based on the separation of armature leakage reactance drop and the armature reaction effect. It gives more accurate results. It makes use of the first two methods to some extent. The zero power factor characteristic is necessary for the calculation by this method. The zero power factor characteristics is the curve of terminal voltage against field current at zero power factor lagging. The phasor diagram under zero power factor lagging condition is shown in Figure 2.93. It will be seen that if the armature resistance drop is neglected.

The leakage reactance voltage drop = $I_a^* X$

Generated e.m.f. E and terminal voltage V are all in phase. Also the three m.m.f. phasors F, F_r and F_a are all in phase, $\overline{E} = \overline{V_t} + \overline{IZ_a}$.

And m.m.f. expression $\overline{F} = \overline{F_r} + \overline{F_a}$.

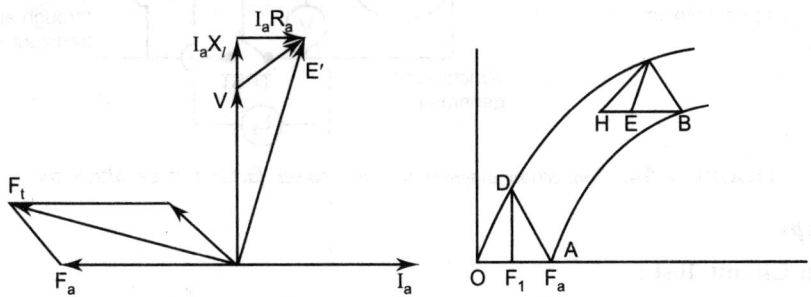

FIGURE 2.93. Phasor diagram and zero power factor characteristics

Figure 2.93 shows the zero power factor characteristic of an alternator. Now obtain the ZPF characteristic. Point A is obtained from a short circuit test with full load armature current. Hence, OA represents field current producing full load current under short circuit (which is naturally at a phase angle of $90°$ lag ($R_a << X_s$)). OA is equal and opposite to demagnetization armature reaction F_a and also to circulate the current through the leakage reactance which is required to induce e.m.f .

$$OA = F_a + F_x = F_{sc}$$

Where F_x – field excitation corresponding to the emf induced in the leakage reactance (X). The triangle OAD must everywhere fit between the two characteristics. It follows that the ZPF curve should be the OCC shifted by the distance $AD = BH$. The triangle OAD is called Potier triangle.

From *OCC* and *ZPF* characteristics we can measure no load voltage and load voltage. Then we can easily calculate the value of regulation.

Equipments and Components Required

S. No.	Name	Specifications	No.
1.	Alternator	Rated voltage =	1
		Rated current =	
		Rated hp =	
		No. of pole =	
		Rated frequency	
		No. of phases =	
2.	Voltmeter		1
3.	Ammeter		1
4.	Rehostate		1
5.	Tachometer		1

Connection Diagram

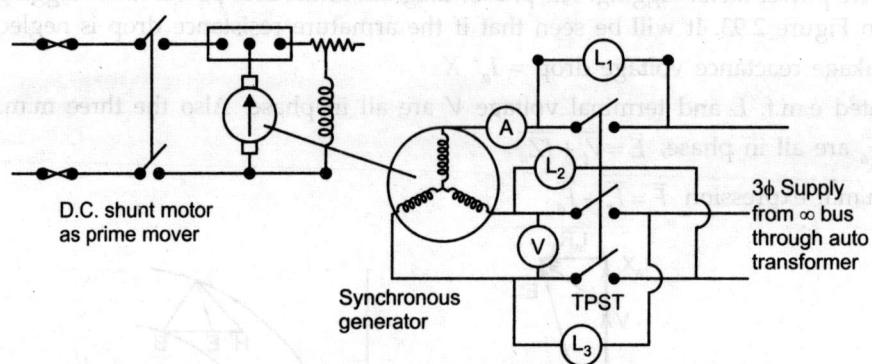

D.C. shunt motor as prime mover

Synchronous generator

TPST

3φ Supply from ∞ bus through auto transformer

FIGURE 2.94. *Connection diagram for zero power factor test on alternator*

Procedural Steps

(A) Open Circuit Test :

1. Make the connection diagram as shown in Figure 2.94.
2. Keep TPST switch in open position.
3. Start D.C. motor and adjust its speed to rated speed (Synchronous speed) and keep it constant through out the experiment.
4. Now increase field current form zero until saturation region occurs and note down field current and corresponding open circuit voltage. While taking the readings measure the speed by tachometer of the set and keep it constant.

(B) Zero Power Factor Test :

1. Make the connection as shown in Figure 2.94.
2. Start the D.C. motor and bring it to synchronous speed.
3. Adjust the field current of alternator to get the rated terminal voltage.

4. Synchronise the alternator to the main through autotransformer.

5. By keeping speed constant, adjust the alternator field such that the alternator delivers the full load current at zero power factor (means your wattmeter reads zero).

6. Repeat the experiment for different voltage settings. Change the voltage of main supply side by autotransformer and the alternator voltage is changed by alternator field and then repeat the steps from (5) to (7).

Observation

(A) Open Circuit Test :

At constant speed .... rpm.

S. No.	Field current (Amp)	Terminal voltage V_o (volts)
1.		
2.		
3.		
4.		
5.		

(B) Zero Power Factor Test :

At constant armature current .... Amp

S. No.	Field current (Amp)	Terminal voltage V_o (volts)
1.		
2.		
3.		
4.		
5.		

Calculation

1. Plot the open circuit and zero power factor characteristics on the same graph paper.
2. Sketch the Potier triangle on the graph paper.
3. Determine regulation at 0.8 lagging , 0.8 leading and unity power factor.
4. Compare the results with the results obtain in the previous experiment.

Results

Percentage regulation in E.M.F. method = ...

Percentage regulation in M.M.F. Method = ...

Percentage regulation in zero power factor Method = ...

2.6.7 Experimental Quiz

1. Why does the armature terminal voltage change as the synchronous generator is loaded?
2. What happens to the value of the synchronous reactance X_s, if air gap is increased?
3. For zero power factor lagging load on an alternator, armature mmf F_a assist F. Why?

4. In an alternator, a lagging current has the effect of weakening the field, but in a synchronous motor, the effect of lagging current is to strengthen the main field. Justify the statement.

5. Why are alternators rated in kVA? What is the necessity of mentioning the power factor at their name plate?

6. A turbo alternator synchronized with an infinite bus system is supplying its rated power. What happens to the alternator if its field circuit gets open circuited accidentally?

7. A turbo alternator synchronized with an infinite bus system operates at unity power factor. If its field current is reduced to 80% of its previous value, will it be absorbing or delivering reactive power?

8. A three phase alternator is operating at 0.8 power factor lagging with respect to the excitation voltage the nature of armature reaction mmf produced by armature current is magnetizing / demagnetizing / partly cross magnetizing and partly demagnetizing / partly magnetizing and partly cross magnetizing?

9. The balanced short circuit current of a poly phase alternator is 20 Amps at speed 1000 rpm. What will be the short circuit current for the same field current at 900 rpm?

10. A synchronous generator connected to the infinite bus bar is working at half load. If an increase in its field current causes a reduction in the armature current, then the generator is deliver / absorb reactive power to / form the bus?

EXPERIMENT NO. 15

2.6.8 To Determine Experimentally the Direct Axis Reactance (X_d) and Quadrature Axis Reactance (X_q) of Synchronous Machine

Apparatus

Ammeter : 0 – 15 A; 0 - 2.5A.

Voltmeter : 0 – 40 V; 0 – 250V.

Wattmeter : 0 -1400 W.

Procedure

(a) Sip Test

1. Check the phase sequence and voltage of the synchronous machine and the bus bar.

2. Remove the field excitation and connect RYB terminals of machine to RYB of bus bar through a three phase variac.

3. Start DC motor and adjust its speed near to synchronous speed, switch on the AC supply and apply a small voltage. The voltmeter connected across field winding should fluctuate and remains within the rated value, if the slip is small and direction of rotation is correct. The armature current should also fluctuate.

4. Increase the AC voltage applied to the armature from the variac, such that maximum current is nearly equal to the rated full load current of the armature.

5. Note down the maximum and minimum value of armature applied voltage and current.

6. Calculate the applied voltage per phase $(V_P = V_L / \sqrt{3})$.

7. Record the readings for different applied voltage.

(b) Maximum lagging current test for determination of X_q:

1. First run the synchronous machine as an alternator with the help of prime mover (dc machine) and synchronize it with infinite bus bar.

2. Switch off the DC supply to the motor so that the synchronous machine will run as a synchronous motor.

3. Gradually reduce the excitation to zero.

4. Reverse the field connection with the help of DPDT switch.

5. Increase the excitation slowly in the negative direction till the machine shows sign of falling out of step. Note this field current, also note the line current.

Static test for determination of Sub-transient reactance

1. With the rotor of the synchronous machine standstill apply single phase voltage from a variac across two phase winding connected in series. The field is short circuited through an ammeter.

2. Rotate the rotor manually such that two positions of flux linking are apparent from the induced field current.

3. Let the armature voltage and current corresponding to the minimum induced field current be V_q and I_q'' and armature voltage and current corresponding to the maximum induced field current be V_d and I_d''.

4. Record the readings for different applied voltages.

For slip test

$$Z_d - \text{maximum phase voltage/minimum phase current}$$

$$X_d = \sqrt{(Z_d^2 - Re^2)}$$

$$Z_q - \text{minimum phase voltage/maximum phase current}$$

$$X_q = \sqrt{(Zq^2 - Re^2)}$$

$$Re - \text{Equivalent resistance /Phase of the alternator}$$

S. No.	Voltage		Current				
	Max	Min	Max	Min	Xd	Zq	Speed
1.							
2.							
3.							
...							

Static Test for X_d'' and X_q''

$$Z_d'' = V_d / (2I_d'')$$
$$Z_q'' = V_q / (2I_q'')$$

$$X_d'' = \sqrt{(Z_d''^2 - Re^2)}$$
$$X_q'' = \sqrt{(Z_q''^2 - Re^2)}$$

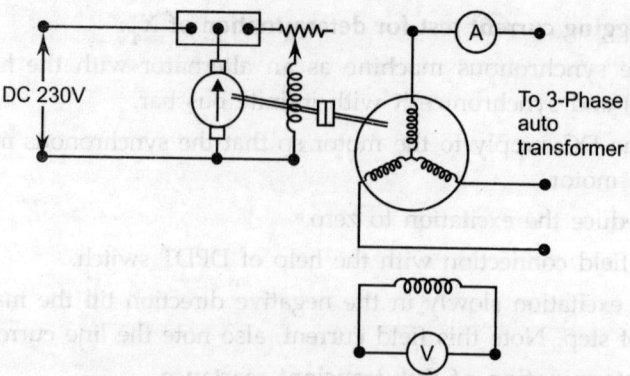

FIGURE 2.95. *Circuit diagram for Slip Test*

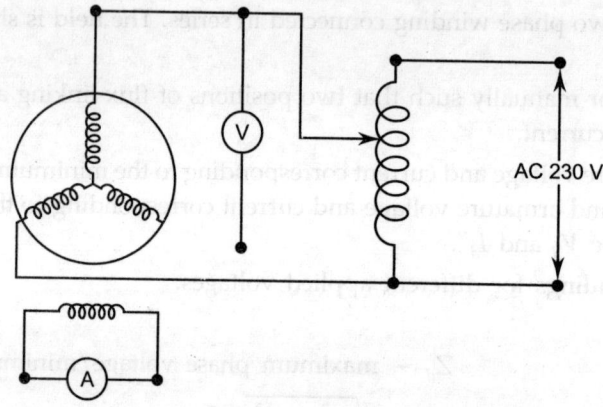

FIGURE 2.96. *Circuit Diagram for Static Test*

Short Answer Type Questions

1. Define direct axis and quadrature axis reactance of a synchronous machine?
2. What do you mean by static test of a synchronous machine?

EXPERIMENT NO. 16

2.6.9 Determination of Positive, Negative and Zero Reactance of Synchronous Machine

Apparatus

Ammeter : 0 – 15 A; 0 – 2.5A.

Voltmeter : 0 – 40 V; 0 – 250V.

Wattmeter : 0 – 1400 W.

Determination of X_1 (Positve Sequence)

A system component operating under balanced condition of voltage and current is in effect in a positive sequence mode. The positive sequence reactance X_1 of a synchronous machine under steady state condition in the direct axis synchronous reactance X_d of the machine. The positive quence impedance can also be defined as the impedance offered by the machine to the flow of sitive sequence currents in the armature windings create a magnetic field that rotates in the rmal direction in the air gap.

$$X_A = E/I_{SC}$$

1. **Open Circuit Test :**

   (a) Run the machine at rated speed after making connections as shown in Figure 2.97(a).

   (b) Connect a voltmeter and ammeter according to the circuit diagram.

   (c) Note the reading at different exciting current.

S. No.	Voltage	Field current	Speed
1.			
2.			
3.			
...			

2. **Short Circuit Test :**

   (a) Run the machine at rated speed after connecting it as shown in Figure 2.97(b).

   (b) Apply low voltage to the field circuit so that exciting current is small. Alternately connect a high resistance rheostat in the field circuit with full field voltage applied connect an armature in the field circuit.

   (c) Apply three-phase short circuit at the synchronous machine terminal with an ammeter connected in any phase.

   (d) Measure the short circuit current corresponding to the field current given by the ammeter reading.

S. No.	Armature current I_A	Field current I_F	Speed
1.			
2.			
3.			
...			

termination of X_2 (Negative Sequence)

e negative sequence reactance X_2 can also be obtained by deriving the machine at rated speed th a low excitation and with a sustained two phase short circuit between the open phase and y short circuited phase be V_{OS} and the short circuit current I_{SC}. If a voltmeter is connected with current coil excited by I_{SC} and voltage coil by V_{OS} it measures the negative sequence impedance.

$$Z_2 = V_{OS}/3 \cdot I_{SC}$$
$$X_2 = Z_2 \sin \varphi, \text{ where } \varphi = \cos^{-1}(P/V_{SC} \cdot I_{SC})$$

(a) Run the machine at rated speed.

(b) Short circuit two phases of the alternator through an ammeter and the current coil of the wattmeter as shown in Figure 2.97(c).

(c) Connect the voltage coil of the wattmeter and the voltmeter between the open phase and any short circuited phase.

(d) Gradually increase the excitation such that the short circuit current does not exceed its full load value.

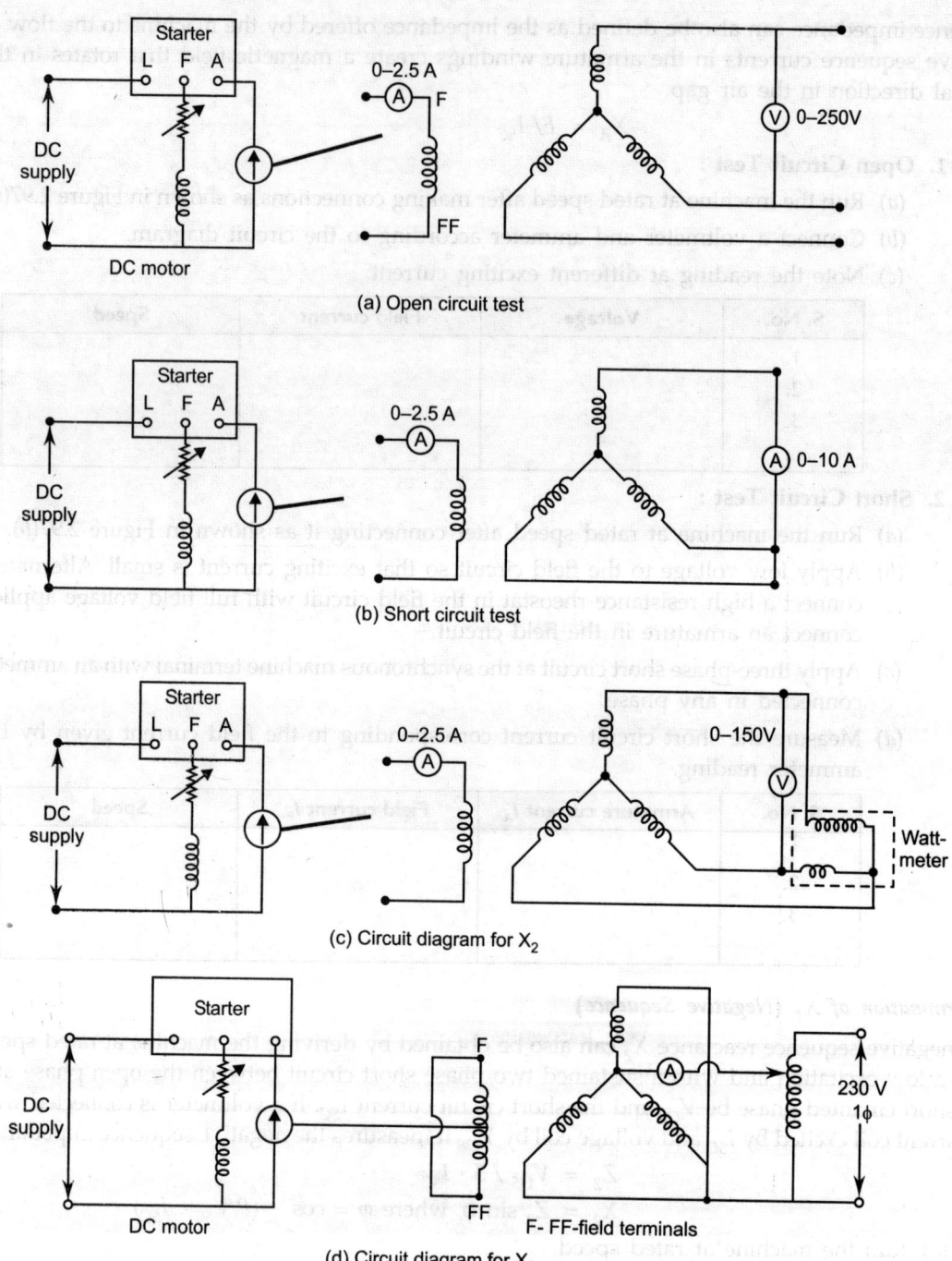

(a) Open circuit test

(b) Short circuit test

(c) Circuit diagram for X_2

(d) Circuit diagram for X_0

F- FF-field terminals

FIGURE 2.97

(e) Take reading of voltage, current and power.

S. No.	V_{0s}	I_{sc}	Power	Wattmeter reading
1.				
2.				
3.				
...				

termination of X_0 (Zero Sequence)

e machine is driven at rated speed. All phases are connected in parallel and connect the ltmeter and ammeter according to the circuit diagram shown in Figure 2.97(d).

(a) Connect the armature winding in parallel according to the circuit diagram.

(b) Run the machine at rated speed.

(c) Apply low voltage from a variac and measure both voltage and current taken by the armature windings

$$X_o = 3 \cdot V_0 / I_0$$

S. No.	V_0	I_0	Speed
1.			
2.			
3.			
...			

EXPERIMENT NO. 17

.10 To Study the Power Angle Characteristic of Synchronous Machine Under Following Conditions :

(a) Variable load with constant excitation and also observe the steady state stability limit.

(b) Varying excitation with constant load.

eory

ıder steady state operation of a synchronous machine when connected to an infinite bus, the ase angle between the no-load induced emf and the infinite bus voltage is the power angle δ. machine armature resistance is neglected (because in large machines $R_a \ll X_d$ or X_q). The ation between power P, excitation emf E and power angle δ under steady state condition for ient pole machine is given by

$$P = \frac{EV}{X_{dt}} \sin\delta + \frac{V^2}{2}\left(\frac{1}{X_{dt}} - \frac{1}{X_{qt}}\right)\sin(2\delta)$$

Where $X_{dt} = X_d + X_e$ and $X_{qt} = X_q + X_e$, X_e – external reactance.

For a cylindrical rotor machine (turbo-alternator) $X_d = X_q$ and the second term in the above pression will be absent. Under transient conditions, the field flux linkage is assumed to remain ıstant and with such as assumption, the relationship between power and angle δ is given by

$$P = \frac{E'V}{X'_{dt}} \sin \delta - \frac{V^2}{2}\left(\frac{1}{X_{dt}} - \frac{1}{X_{qt}}\right) \sin (2\delta)$$

Where E' is the voltage behind transient reactance along direct axis

$$X'_{dt} = X'_d + X_e$$

When the alternator is connected to infinite - bus, which has hundreds of generator and voltage and frequency are constant, what ever be the load on it. Under ideal situations infinite bus could supply infinite-power, but practically it is not so. The power (P and Q) of given alternator can vary by :

    (*i*) The change in excitation current

    (*ii*) The change in mechanical power input to the machine.

Effect of varying excitation current of alternator when connected to infinite-bus

After synchronization of alternator to a infinite-bus the induced voltage E_o is equal to and phase with the terminal voltage of the system. Although the generator is connected to the system it delivers no power; it is said to float on the line.

Now, if excitation current is increased means the alternator is overexcited, the generator voltage E_o will increase and alternator starts delivering reactive power to the system. In this mode of operation synchronous machines is called rotating condenser. Let us now decrease the excitation current (under excited condition), the generated emf will be less than system voltage and alternator draws reactive power. This reactive power produces part of the magnetic field required by machine and rest is supplied by excitation current.

Effect of varying mechanical power of alternator when connected to infinite-bus

Let us consider that the alternator is floating on line and the mechanical power is increased through the prime mover input (*i.e.*, the steam of steam turbine). The rotor will accelerate and will slip ahead of phasor of system voltage, leading it by a phase angle δ. A current will flow as the generator feeds active power into the system.

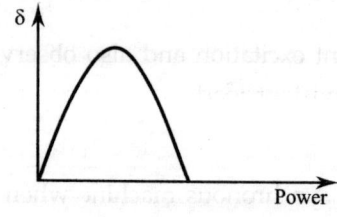

FIGURE 2.97(A)

Apparatus

    Ammeter : 0 – 15 A; 0 – 2.5A.

    Voltmeter : 0 – 40 V ; 0 – 250V.

    Wattmeter : 0 –1400 W

    Motor – Generator Set

    Stroboscope

Procedure

    1. Make the connections as shown in Figure 2.98.

2. Start the DC motor and synchronize the alternator with the infinite bus using any one of the standard methods.

3. Adjust the motor field current in case the wattmeter does not read zero, observe the position of the pointer fixed to the alternator shaft using a stroboscope synchronized to the line frequency with the alternator output zero.

4. Increase the output of the alternator by decreasing the DC motor field current. Note down the movement of the pointer as observed using the stroboscope. The angle through which the pointer has moved from the position corresponding to zero output will be δ mechanical degree.

5. Note down the wattmeter reading.

6. Gradually increase the alternator power output and record the value of δ for various values of power. Note the steady state power output and power angle limits.

7. Plot a graph between P and δ.

8. Repeat the above experiment for different values of excitation.

Observations

(a) E = Constant ($I_f = I_{f1}$)

S. No.	P	Δ
1.		
2.		
3.		
...		

(b) E = Constant ($I_f = I_{f2}$)

S. No.	P	δ
1.		
2.		
3.		
...		

(c) $P = P_1$

S. No.	E_f	δ
1.		
2.		
3.		
...		

(d) $P = P_2$

S. No.	E_f	δ
1.		
2.		
3.		
...		

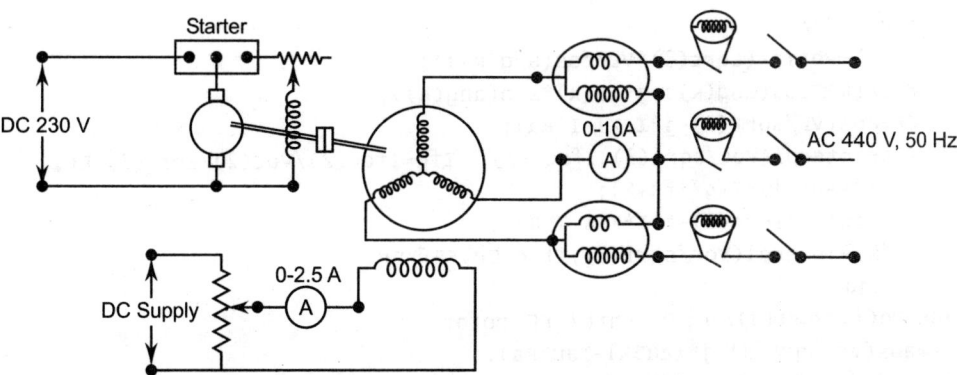

FIGURE 2.98. *Connection diagram for power angle characteristic*

PROBLEM. *Write a Matlab program for calculating the data for v-curves of synchronous machine in motor mode and plot them.*

2.6.11   Matlab Program for Drawing V-curves

```
%***************************************************************
%
% syn_vcurves.m - Plots V-curves for synchronous motor.
%
%***************************************************************
clear all;
VL=2400; KVA=10000; PF=0.8; PR=KVA*PF; % Rated values
Xsu=VL^2/(KVA*1000); % unsaturated synchronous reactance
Xl=0.1*Xsu; % Leakage reactance, ohm
Ra=1e-5*VL^2/KVA; % Armature Resistance, ohm

% OC sat curve data using line voltage
Voc=[0 1700 1950 2200 2300 2400 2500 2600 2700 2800 2900 ...
    3000 3100 3300]';
Ifoc=[ 0 70 83 98 105 115 126 140 160 180 205 235 280 400]';

% Set plot for rated Ia & If
IaR=1000*KVA/VL/sqrt(3)*(PF+j*sin(acos(PF)));
Er=abs(VL/sqrt(3)-j*IaR*Xl-IaR*Ra);
Ifs=interp1(Voc/sqrt(3),Ifoc,Er);
Ifg=Ifoc(2)/(Voc(2)/sqrt(3))*Er;
Xss=(Xsu-Xl)*Ifg/Ifs+Xl; % Saturated Xs
Ef=abs(VL/sqrt(3)-j*IaR*Xss-IaR*Ra);
IfR=interp1(Voc/sqrt(3),Ifoc,Er)*Ef/Er;
axis([0, IfR, 0, abs(IaR)]);

% Set output power values
ncurv=5; Po=linspace(0, PR, ncurv)*1000;

% Set range of PF angle
ang=linspace(-60, 60, 50)*pi/180; n=length(ang);
for i=1:ncurv
    for k=1:n
    Ia(k)=Po(i)/(sqrt(3)*VL*cos(ang(k)));
    I=Ia(k)*cos(ang(k))+j*Ia(k)*sin(ang(k));
    Er=abs(VL/sqrt(3)-j*I*Xl-I*Ra);
    Ifs=interp1(Voc/sqrt(3),Ifoc,Er);  Ifg=Ifoc(2)/Voc(2)/sqrt(3)*Er;
    Xss=(Xsu-Xl)*Ifg/Ifs+Xl;
    Ef=abs(VL/sqrt(3)-j*I*Xss-I*Ra);
    If(k)=interp1(Voc/sqrt(3),Ifoc,Er)*Ef/Er;
      end
Iau=Po(i)/sqrt(3)/VL; % Unity PF point
Er=abs(VL/sqrt(3)-j*Iau*Xl-Iau*Ra);
Ifs=interp1(Voc/sqrt(3),Ifoc,Er);
Ifg=Ifoc(2)/Voc(2)/sqrt(3)*Er;
```

```
Xss=(Xsu-Xl)*Ifg/Ifs+Xl;
Ef=abs(VL/sqrt(3)-j*Iau*Xss-Iau*Ra);
Ifu=interp1(Voc/sqrt(3),Ifoc,Er)*Ef/Er;
Ial=Po(i)/sqrt(3)/VL/0.8; % 0.8 lead PF point
I=Ial*(0.8+j*0.6);
Er=abs(VL/sqrt(3)-j*I*Xl-I*Ra);
Ifs=interp1(Voc/sqrt(3),Ifoc,Er);
Ifg=Ifoc(2)/Voc(2)/sqrt(3)*Er;
Xss=(Xsu-Xl)*Ifg/Ifs+Xl;
Ef=abs(VL/sqrt(3)-j*I*Xss-I*Ra);
Ifl=interp1(Voc/sqrt(3),Ifoc,Er)*Ef/Er;
plot(If,Ia,Ifu,Iau,'o',Ifl,Ial,'*'); grid; hold on;
end
title('V-curves for synchronous motor');
xlabel('Field current, A'); ylabel('Stator current, A');
legend(['KW incr. = ',num2str((Po(2)-Po(1))/1000)], ...
    'Unity PF', '0.8 PF leading',4);
hold off;
%*****************************************************************
```

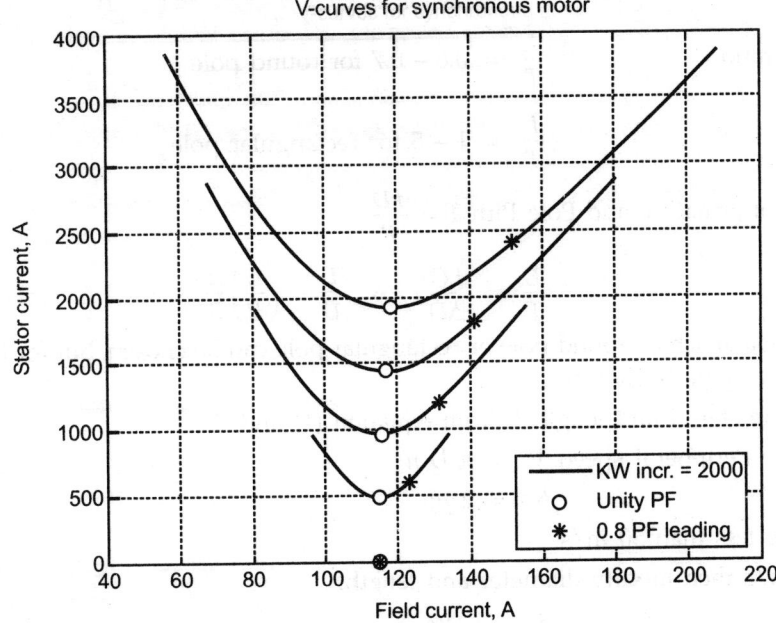

FIGURE 2.99. *Matlab program output*

2.6.12   Design Procedure of Synchronous Machine

2.6.12.1 *Specifications*

kVA = Q =   Voltage, V =

Phase = 3   Frequency, f = 50 Hz

Poles = P Connection = Y/Δ

Type = PF of load = 0.8 Lagging

2.6.12.2 Main Dimensions

Output Equation

$$Q = C_o D^2 L n_s$$
$$Q = \text{kVA Output,}$$
$$K_W = 0.955 = \text{Winding Factor}$$
$$C_o = 11 \, B_{av} \, ac \, Kw \times 10^{-3} = \text{Output Coefficient}$$
$$B_{av} = \text{Specific. Magnetic loading}$$
$$= 0.52 - 0.65 \text{ T for salient pole machine}$$
$$= 0.54 - 0.65 \text{ T for turbo machine}$$
$$ac = \text{Specific Electrical loading}$$
$$= 20.000 - 40,000 \text{ A/M for salient Pole machine}$$
$$= 50,000 - 75,000 \text{ A/M For cylindrical rotor machine}$$

Calculate C_o

Calculate

$$D^2 L = \frac{Q}{(C_o \, n_s)} \qquad \qquad ...(1)$$

Constant

$$Ms = \text{Synchronous speed in rps} = \frac{2f}{P}$$

Choose the ratio

$$\frac{L}{T} = 0.6 - 0.7 \text{ for round pole}$$

$$\frac{L}{T} = 1 - 5 \text{ for rectangular pole}$$

L = Length of armature and Pole Pith $T = \dfrac{\pi D}{P}$

$$\frac{L}{T} = \frac{LO}{AD} \quad \Rightarrow \quad \frac{L}{D} = \left(\frac{L}{P}\right)\frac{\pi}{P} \qquad \qquad ...(2)$$

For salient pole machine round pole or rectangular pole can be chosen, but for turbo machine always-rectangular pole is chosen.

Calculate diameter D and length L from equation (1) and (2)

Calculate Peripheral speed $V_a = \pi \, D \, n_s$

$$Ns = 2 \, f/p$$

V_a should be less than 50 m/s

If V_a = 50 m/s then modify diameter and length,

$$D = \frac{Va}{n * n_s} \qquad n_s = 50 \text{ m/s}$$

$$L = \frac{Q}{Co * D^2 * L * ns}$$

Ducts are used for cooling of stator winding for each 100 mm

Number of ducts $n_d = \dfrac{L}{100}$ in mm

Width of duct w_d = 10 mm = w_d

$$\text{Gross iron length } L_s \ = \ L - n_d \, w_d$$
$$\text{Net iron length } L_i \ = \ org \ Ls$$

2.6.12.3 Starter

For star (Y) Connected stator winding

$$\text{Phase voltage } E_{ph} \ = \ V_L / \sqrt{3}$$

For D connected stator winding

$$E_{ph} \ = \ V_L$$
$$\text{Flux per pole } F \ = \ B_{av} \, T_l$$

$$\text{Turns per phase } T_{ph} \ = \ \frac{E_{ph}}{4.44 \, f \, \phi \, Kw}$$

$$Kw \ = \ 0.955$$

Number of Slots : Select the armature slot pitch y_s as given below

Upto 1.0 kV machine $y_s \ < \ $ mm25 mm

Upto 6.0 kV machine $y_s \ < \ $ mm 40mm

Upto 1.6 kV machine $y_s \ < \ $ mm 60mm

Above 1.5 kV machine $y_s \ < \ $ mm 90mm

$$\text{Slot/Pole /Phase } q \ = \ \frac{\pi D}{3p * y_s}$$

Number of armature slot $S \ = \ 3$

$PQ = $ should be a whole number it not make it.

Modify $q = s/3p$

Total number of armature conductor

$$Z \ = \ 6 \, T_{ph}$$

Conductors per slot $Z_s = z/s$ may be a fraction number of conductors/slot should be an even integer.

Make $Z_s \rightarrow$ even integer

$$\text{Total conductors } z \ = \ Z_s \ x \ s$$
$$\text{Turns per phase } T_{ph} \ = \ Z/6 \text{ (Modified)}$$

Modify the average flux density in the air gap

$$B_{av} \ = \ (B_{av})_{taken} \times (T_{ph})_{old} / (T_{ph})_{modified}$$
$$\text{Modify flux per pole } \Phi \ = \ B_{av} \times t \times L$$
$$\text{Stator slot pitch } y_s \ = \ \pi \, D/S$$

Current per phase

$$I_{ph} \ = \ \frac{kVA \times 1000}{\sqrt{3} \, V_2} \text{ for } Y \text{ connected armature}$$

Minimum tooth width for salient pole machine

$$(w_t) \text{ min } \ = \ \frac{\phi}{0.74 * 1.8 * s * L_i * p}$$

Minimum tooth width for turbo machine

$$(w_t) \text{ min} = \frac{\phi}{1.8 * s * L_i * p}$$

Maximum slot width

$$(W_s) \text{ max} = y_s \times (w_t) \text{ min}$$

Conductor Size

Choosing current density $S_a = 3 - 5 \text{ A/mm}^2$

$$\text{Take } S_a = 4\text{A/ mm2}$$

Area cross section of armature conductor $a_s = \dfrac{I_{ph}}{S_a}$

Keeping in mind W_s max, select the conductor Size.

Slot depth d_s = total copper conductor width + insulation thickness.

Length of mean turn of armature

$$L_{mts} = (2L + 2.5 \, T + 0.6 \text{ kV} + 0.2 \, L) \text{ meter}$$

Stator Resistance

Stator resistance (dc) per phase $r_{dc} = \dfrac{0.021 * T_{ph} * L_{mts}}{a_s}$

where a_s in mm^2 and L_{mts} in meters

DC resistance of conductor in slots (r_{dc}) slot $= \dfrac{0.021 * T_{ph} * 2L}{a_s}$

(a) Copper loss/phase in slots = $1.2 \, I_{ph} \times (r_{dc})$ slot

overhang length of cu conductor = $L_{mts} - 2L$

$$r_{dc} \text{ overhang} = \frac{0.021 * T_{ph} * (L_{mts} - 2L)}{a_s}$$

(b) Copper loss per phase in overhang = $I_{ph}^2 . \, r_{dc}$ overhang

Total copper loss per phase = (a) + (b)

$$r_{ac} = (a + b) / I_{ph}^2$$
$$\text{pu } R_{ac} = I_{ph} \, r_{ac} / E_{ph}$$

Stator Core

Flux in stator core $f_c = f/2$

Flux density in core = $B_c = 1.1. \text{ wb/m}^2$

Depth of stator core $d_c = f_c / (B_c \, x \, L_i)$

$$D_o = D + 2 \, (d_c + d_s)$$

Length of Air Gap

For salient pole Machine

$$SCR = \frac{AT_{fo}}{AT_a} = \frac{\text{No load field mmf}}{\text{Armature mmf/pole}}$$

$$AT_a = \frac{2.7 * I_{ph} * T_{ph} * K_w}{p} \qquad\qquad Kw = .0955$$

$$SCR = 1 - 1.5 \text{ (for salient pole machine)}$$

$$AT_{fo} = \frac{2.7 * I_{ph} * T_{ph} * K_w}{p} * SCR$$

MMF required for air gap = 80% of $AT_{fo} = 0.8 \; AT_{fo}$

$$0.8 \; AT_{fo} = 800,000 \; Bg \; kg \; lg$$

$$l_g = \frac{0.8 * AT_{f0}}{800,000 * B_g * K_g}$$

Assume
$$K_g = 1.15$$

$$K_g = \text{gap contraction factor}$$

$$B_g = B_{av} * \frac{\text{Pole arc}}{\text{Pole pitch}} \qquad (B_g \text{ is generally taken to 0.74})$$

For Turbo Machine

$$SCR = 0.5 - 0.7$$

$$AT_{fo} = SCR \; X \; (\text{armature mmf/pole})$$

$$= SCR * \frac{act}{2}$$

$$0.8 \; AT_{fo} = 800,000 \; B_g \; K_g \; L_g.$$

$$Lg = \frac{0.5 * act}{B_g \; K_g} * 10^{-6} * SCR$$

The value of B_g = 1.5 Bav, Kg = 1.1

Poles : For salient pole machine only

Flux in pole body = leakage factor × ϕ_m

$$\phi_p = 1.2 \; \phi_m$$

take, flux density in pole body B_p = 1.6 wb/m^2

Area of pole body $A_p = \dfrac{\phi_p}{B_p}$

Width of pole $b_p = \dfrac{A_p}{0.98 * L}$ for rectangular pole

Width of pole $b_p = \dfrac{A_p}{\pi / 4}$ for round pole.

For turbo machine (use rectangular pole)

$$A_p = \frac{\phi_p}{B_p} = \frac{1.2 \; \phi_m}{1.6}$$

Pole width $b_p = \dfrac{A_p}{0.98 * L}$

Damper Winding : Total cross sectional area of damper bars

$$A_d = \frac{0.2 * act}{S_d} \; ; \; S_d = 4 \text{ A/mm}^2$$

Pole Yoke Design

depth of pole yoke $D_y = \dfrac{A_y}{0.98 * L}$ where A_y – yoke Area

$$A_y = \frac{\phi_m / 2}{1.2} = \frac{\phi_m}{2.4}$$

(Assume flux density in pole yoke = 1.2T)

Diameter of shaft $d_{shaft} = 0.1 \dfrac{Kw}{R_{fm}}$ meters

Number of damper bars $N_d = \dfrac{\text{Pole arc}}{Y_{sd}}$

$$N_d = \frac{B_p}{0.8 \, Y_{sd} Ad} \text{ (integer)}$$

Sectional area of each damper bar $a_d = \dfrac{A_d}{N_d}$

In case of circular damper bar $a_d = \dfrac{\pi}{4} d_d^2$

Diameter of each damper bar $d_d = \sqrt{\dfrac{4 * a_d}{\pi}}$

Length of damper bar = 1.1 L

$$A_{ring} = (0.8 - 1)Ad$$

For salient pole machine $h_s = 2d_d$

Air gap in the pole shoe region remains non-uniform to make the flux density wave sinusoidal. The air gap length varies as $L_g = 1.5 \, l_g$ (generally taken)

$$H_1 = AB = OC - CB - AO$$

Pole Core Design

Height of pole core

Area of the pole core $A_{pc} = \phi_m / 1.2$

$$A_{pc} = L_i \times W_{pc} = 0.98L \, W_{pc}$$

Width of the pole core $W_{pc} = \phi_m / (1.2 \times 0.98 \times L)$

Design of Field Winding

Length of mean turn of field winding $L_{mtf} = 2L + p \, (b_p + d_f + .01)$

$$L \rightarrow \text{meters}$$
$$B_p \rightarrow \text{meters}$$
$$d_f \rightarrow \text{meters}$$

Select field winding depth d_f

Pole pitch (Meters)	Winding depth d_f (mm)
0.1	25
0.2	35
0.4	40

1. Select exciter voltage 250 V for field system. $V_e = 250$ V

   Voltage across each field coil $E_f = \dfrac{V_e}{P}$

2. Height of the field winding

$$h_f = hpl - \text{space taken for spools and flanges (20mm)}$$

3. Calculate the full load field mmf at 0.8 pf lag of load,

$$OA = AT_{fo}$$
$$AB = AT_a$$
$$A_c = K_r\, AB.$$

Find
$$OD = AT_{fl}$$
$$= \text{full load field mmf}$$
$$K_r = \text{From Figure 2.100}$$
$$= \text{cross reaction coefficient}$$

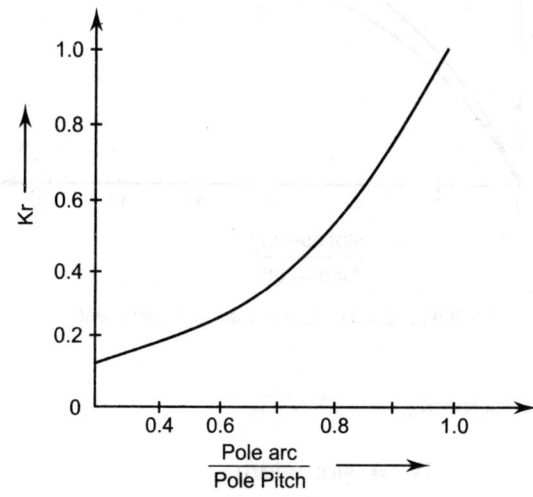

FIGURE 2.100. *Cross reaction coefficient*

$$AT_{ft} = 2\, AT_a \rightarrow \text{for turbo machines}$$

4. Area of field conductor

$$a_f = \frac{AT_{fl}\, P\, L\, mtf}{Ef} \qquad \begin{array}{l} p = 0.021 \\ a_f = mm^2 \end{array}$$

5. Field current $I_f = a_f \cdot s_f = 3 - 4$ A/mm^2

6. Field turn $T_f = AT_{fl}\,/I_f$

7. Resistance of field winding $R_f = \dfrac{P * Lmt * T_f}{A_f}$

8. Copper loss in each field coil $Q_f = I_f^2 R_f$

Dissipating surface of coil $S = 2L_{mt} (h_f + d_f)$

$$\text{Cooling coefficient } C_f = \frac{0.08}{(1+0.1 V_a)}, V_a = \frac{\pi DN}{60}$$

$$\text{Temperature rise } Q = Q_f \frac{C_f}{S}$$

The temp rise should be with in specified limit (50°C) If it increases or exceeds the specified limit then increase the depth of the winding, this will increase the a_f, thus reaucing R_f and $I_f^2 R_f$.

Magnetic Circuit

(i) MMF For air gap

Calculate slot width /gap length corresponding to this ratio the craters coefficient K_{cs} is chosen, (Figure 2.101) for open slots.

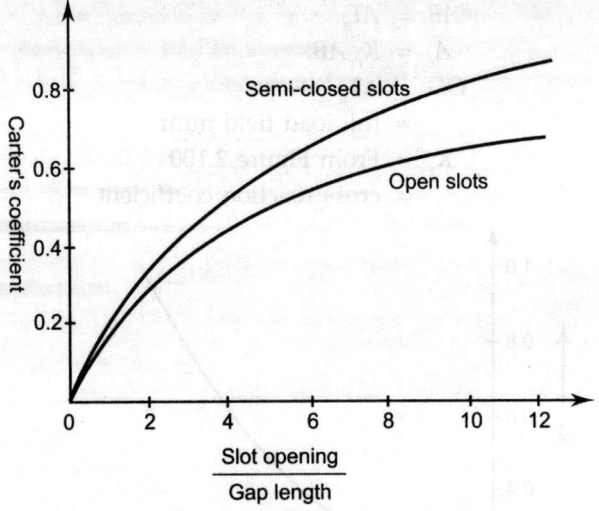

FIGURE 2.101. *Carter's air gap coefficient*

$$\text{Gap contraction factor } K_{gs} = \frac{Y_s}{Y_s - k_{cs} W_s}$$

$$W_s = \text{slot width}$$

MMF required for air gap $AT_g = 800,000 B_g K_{gs} l_g.$

(ii) MMF For stator teeth

Width of teeth at $1/3^{rd}$ height from narrow end

$$W_{t1/3} = \frac{\pi\left(D + \frac{2}{3}d_s\right)LW_s}{\phi_m}$$

$$\text{Flux density in teeth } B_{t1/3} = \frac{\phi_m}{0.74 * \frac{s}{p} * W_{t1/3} * L_i}$$

Corresponding to this flux density find mmf/m, at_t from Figure 2.102.

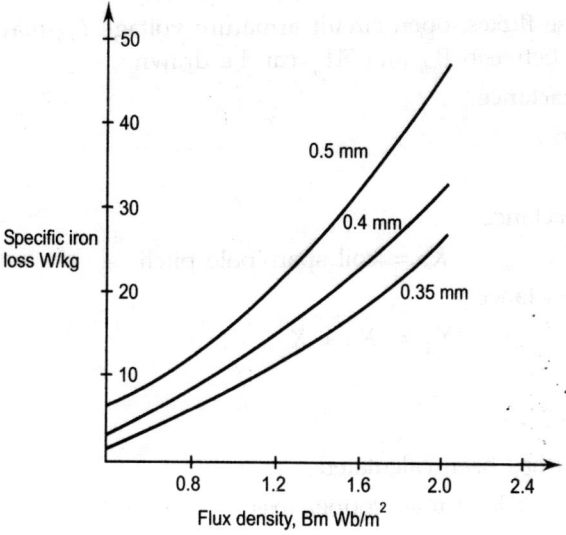

FIGURE 2.102. *Specific iron loss for rotating machines*

Total mmf required for teeth

$$AT_t = at_t \times \text{Total Slot depth in } m$$
$$= at_t \times ds$$

(iii) MMF FOR CORE

$$\text{Area of core } A_c = d_c \times L_i$$
$$\text{Flux density in core } B_c = \varphi/2A_c$$
$$\text{MMF per meter} = atc \text{ (refer Figure 2.102)}$$
$$\text{Length of flux path in core } L_c = \pi(D + 2d_s + d_c)/2p$$
$$\text{MMF for core } AT_c = at_c \times l_c$$
$$AT_2 = AT_g + AT_c$$

(iv) MMF for poles

$$\text{Width of pole shoe} = b_s$$

$$\text{Distance between adjacent pole shoe } CS = \frac{\pi(D_r - h_1)}{P} - BS$$

$$D_r = D - 2l_g$$
$$\text{Height of pole shoe at pole tip} = h_s$$
$$\text{Width of pole body} = b_p$$

Distance between bodies of adjacent poles

$$C_p = \pi(D_r - H_1 - hpl)/p - b_p$$

Leakage flux from poles

$$\varphi_{sl} = 4\mu \ Atl \ [(l_s \ h_s + 1.47 \ h_s \ \log10 \ (1 + 2^\pi \ cs^{bs})]$$
$$L_s = \text{Gross iron length}$$

Leakage flux between pole bodies

$$\varphi_{pl} = 2 \ \mu \ AT_l \ [LPHP1]$$

Here we have calculated No-load mmf AT_{fo} for one value of flux φ. Similarly we can also calculate the values of At_{fo} at different values of φ i.e., 1.1. φ, 1.2 φ, 1.3 φ etc.

Corresponding to these fluxes, open circuit armature voltage E_{ph} may be determined. Now open circuit characteristic between E_{ph} and AT_{fo} can be drawn.

Armature Leakage reactance

Specific slot permenance

Slot leakage reactance

Over hang leakage reactance

$$K_s = \text{coil span/pole pitch}$$

Total stator leakage reactance

$$X_{sl} = X_{ss} + X_o$$

pu leakage reactance

LOSSES

(i) Stator copper loss has been calculated.

(ii) Stray load losses = 20% of total copper loss

(iii) Iron loss

   (a) Stator teeth

   Tooth width at middle of stator tooth

   $W_t = \dots\dots\dots\dots\dots\dots\dots\ w_x$

   Weight of stator teeth = $w_x * L_i * S^* d_s * 7.8 * 10^3 \ k_g$ meters

   Figure 2.102 find specific iron losses

   Corresponding to $B_{t\ 1/3}$

   Iron losses in teeth = weight X_{sp} iron loss.

   (b) Armature core (Stator)

   Mean core diameter = $D + 2d_s + d_c$

   Weight of stator core = $(D + 2d_s + d_c) \times d_c \times 7.8 \times 10^3$

   Corresponding to flux density in core B_c, find specific iron loss from Figure 3.

   Iron loss in core = weight × watt/kg

   Total Iron loss= iron loss in core + iron loss in teeth.

(iv) Friction and windage losses = 0.7% of kVA rating

(v) Field copper loss = $p^* I_f^2 \ R_f$

   Total losses = copper loss in stator + stray load loss + Total iron loss in stator

   $$+ \text{ friction and windage loss} + \text{field copper losses}$$

$$\text{rated output} = \text{kVA} \times \text{PF}$$

$$\text{Input} = \text{output} + \text{losses}$$

STATOR TEMPERATURE RISE

Cooling coefficient for outer surface = 0.03

Area of outer surface = $\pi \ DL \ m^2$

Loss dissipation form outer surface = $\pi \ DL \ /0.03$ Watt/°C

Inner surface area = πDL

Cooling coefficient for inner surface = $0.03 / 1 + 0.1\ V_a$

$$Va = \frac{\pi DN}{60} \text{ where } N = \text{rpm}$$

Loss dissipation from inner surface = $\pi\ DL$

2.6.13 Review Exercises

I. Answer in Brief :

1. What is the maximum rating of alternators for generating electricity in India?
2. Mention two the differences between hydro generators and turbo generators.
3. Mention the three advantages of the Chorded winding.
4. Mention at least three methods for synchronization.
5. Draw open circuit and short circuit characteristics of a synchronous machine.
6. Show the effect of speed on OCC of synchronous machine.
7. Define voltage regulation of an alternator.
8. Name the three methods for finding voltage regulation of an alternator.

II. Multiple Choice Questions :

1. The maximum speed of the prime mover in case of a turbo synchronous generator will be :
   (a) 750 rpm   (b) 1500 rpm   (c) 3000 rpm   (d) 6000 rpm.
2. An exciter is nothing but a :
   (a) dc shunt generator   (b) dc series generator
   (c) dc compound generator   (d) dc shunt motor.
3. An alternator coupled to which prime mover will usually have the highest rotating speed :
   (a) steam engine   (b) reciprocating diesel engine
   (c) francis turbine   (d) steam turbine.
4. Salient pole type alternator are generally used on :
   (a) low voltage alternator   (b) hydrogen cooled prime mover
   (c) high speed prime-mover   (d) low and medium speed prime-mover.
5. Which of the prime mover is least efficient :
   (a) gas turbine   (b) petrol engine
   (c) diesel engine   (d) steam engine.
6. In which coil the harmonic component of the generated emf will be more :
   (a) full pitch coil   (b) short pitch coil
   (c) long pitch coil   (d) same in all coils.
7. In case of turbo alternators the rotor is usually made of :
   (a) cast iron   (b) forged steel
   (c) laminated stainless steel   (d) manganese steel.
8. In case of alternator which supplies resistive load :
   (a) the armature reaction flux will be almost zero
   (b) the armature reaction flux will be along the axis of the field
   (c) the armature reaction flux will be at 90° with the field axis
   (d) the armature reaction flux will be at 180° with the field axis.

9. A 12 pole alternator will pass through how many elect. degrees in one complete revolution :

(a) 60 deg.   (b) 360 deg.   (c) 1080 deg.   (d) 2160 deg.

10. Which harmonic is totally eliminated in an alternator by using a fractional pitch of 4/5:

(a) third   (b) fifth   (c) seventh   (d) ninth.

11. Three phase alternator are usually star connected :

(a) to save copper   (b) to reduce conductor size

(c) to obtain higher terminal voltage   (d) to reduce windage losses.

12. Damper winding on alternator results in all of the following except:

(a) increases instability of machine

(b) elimination of harmonic effect

(c) absorption of energy of oscillations when operating in parallel with other alternators

(d) suppression of spontaneous hunting when supplying power to transmission line with high resistance to reactance ratio.

13. Short circuit ratio of turbo alternator is usually :

(a) 0.5   (b) 0.5 to 0.7   (c) 0.1 to 1.7   (d) 2 to 3

14. The speed of an alternator is changed from 3000 rpm to 1500 rpm. The generated emf/phase will become :

(a) one fourth   (b) half   (c) double   (d) unchanged.

15. Zero power factor method is generally used to determine :

(a) synchronous impedance of alternator   (b) efficiency

(c) voltage regulation   (d) none of the above.

16. The effect of cross magnetizing armature reaction in an alternator can be reduced by :

(a) using interlopes   (b) improving cooling

(c) shifting brush position   (d) reduce rotor speed.

17. Which of the following has significant influence on the power factor of an alternator :

(a) load connected   (b) excitation

(c) HP of prime-mover   (d) speed of alternator.

18. When the power factor of load is unity, the armature flux of an alternator will be -

(a) demagnetizing   (b) cross magnetizing

(c) square waveform   (d) in phase with current.

19. Under which of the following conditions hunting of synchronous motor is likely to occur :

(a) periodic variation of load   (b) over-excitation

(c) overload for long time   (d) small and constant load.

20. Generally the three phase synchronous motor will have:

(a) no slip rings   (b) one slip ring   (c) two slip rings   (d) three slip rings

21. When the excitation of an unloaded salient pole synchronous motor suddenly gets disconnected :

(a) the motor stops

(b) it runs as a reluctance motor at the same speed.

(c) it runs as a reluctance motor at the lower speed.

(d) none of the above.

22. The phase sequence of voltage generated in the alternator can be reversed by reversing its field current :

(a) True   (b) False

23. To obtain the sinusoidal voltage the poles should have the shape such that the length of the air gap at any point is

    (a) Proportional to $1/\sin\theta$, where θ in electric degrees between the point in the question and centre of the pole.

    (b) Proportional to $1/\cos\theta$

    (c) Proportional to $\sin\theta$

    (d) Proportional to $\cos\theta$.

24. The armature reaction of an alternator will be completely magnetizing when

    (a) Load power factor is unity   (b) Load power factor is zero lagging

    (c) Load power factor is zero leading   (d) Load power factor is 0.7 lagging.

25. When the alternators are running in proper synchronism the synchronizing power will be zero -

    (a) True   (b) False.

26. If the input to the prime mover of an alternator is kept constant but the excitation is changed then

    (a) kVAR output is changed   (b) power factor of the load remain constant

    (c) kVA output is changed   (d) (a) or (b) both.

27. The coupling angle or load angle of synchronous motor is defined as

    (a) the angle between the rotor and stator pole of same polarity.

    (b) the angle between the rotor and stator pole of opposite polarity.

    (c) the angle between the rotor and stator teeth.

    (d) none of the above.

28. The alternators are normally designed for the torque angle of the order of

    (a) 3 deg. to 15 deg.   (b) 2 rad to 3 rad.

    (c) 15 deg. to 30 deg.   (d) 1 deg to 3 deg.

29. The armature reaction of an alternator influences

    (a) Windage losses   (b) Operating speed

    (c) Generated voltage per phase   (d) Waveform of voltage generated.

30. In synchronous alternators which of the following coil will have emf closer to sine wave form

    (a) Concentrated winding in full pitch coils

    (b) Concentrated winding in short pitch coils

    (c) Distributed winding in full pitch coils

    (d) Distributed winding in short pitch coils.

31. The maximum power developed in the synchronous motor will depend upon :

    (a) Rotor excitation only

    (b) Maximum value of coupling angle

    (c) Supply voltage only

    (d) Rotor excitation supply voltage and maximum value of coupling angle.

32. A three phase 4-pole, 24 slots alternator has its armature coil short pitched by one slot. The distribution factor of alternator will be

    (a) 0.96   (b) 0.9   (c) 0.933   (d) 0.966.

33. When the alternator are running in the proper synchronism the synchronizing power will be

    (a) zero   (b) maximum   (c) minimum   (d) none.

34. The coupling angle or load angle of synchronous motor is
    (*a*) Angle between the rotor and stator poles of same polarity
    (*b*) Angle between the rotor and stator poles of opposite polarity
    (*c*) Angle between the rotor and stator teeth
    (*d*) none.

35. If the armature current leads E_f by 90 degrees, the armature reaction mmf is completely
    (*a*) magnetizing   (*b*) demagnetizing
    (*c*) cross magnetizing   (*d*) none.

36. Simple synchronous motors are not self starting because
    (*a*) the direction of rotation is not fixed
    (*b*) the direction of instantaneous torque reverses after half cycle.
    (*c*) starter cannot be used on these machines.
    (*d*) starter winding is not provided on these machines.

37. In case one phase of a three phase synchronous motor is short circuited, the motor will
    (*a*) not start   (*b*) run at half of synchronous speed
    (*c*) run with excessive vibrations   (*d*) take less than the rated load.

38. A synchronous motor can develop synchronous torque
    (*a*) when motor is under loaded   (*b*) while motor is overexcited
    (*c*) only at synchronous speed   (*d*) below or above synchronous speed.

39. As compared to an induction motor of the same size the air gap in synchronous motor is
    (*a*) 3 to 5 times   (*b*) same
    (*c*) 0.5 to 3.5 times   (*d*) less than 0.5.

40. The power developed by a synchronous motor will be maximum when the load angle is
    (*a*) zero   (*b*) 45 degrees   (*c*) 90 degrees   (*d*) 120 degrees.

41. A synchronous motor is running on a load with normal excitation. Now if the load on the motor is increased
    (*a*) power factor as well as armature current will decrease
    (*b*) power factor as well as armature current will increase
    (*c*) power factor will increase but armature current will decrease
    (*d*) power factor will decrease but armature current will increase.

42. The back emf of a synchronous motor depends on
    (*a*) speed   (*b*) load   (*c*) load angle   (*d*) none.

43. A synchronous motor can be made self starting by providing
    (*a*) damper winding on rotor poles
    (*b*) damper winding on stator
    (*c*) damper winding on both rotor as well as on stator
    (*d*) none of the above.

44. The speed regulation of synchronous motor is always
    (*a*) 1%   (*b*) 5%
    (*c*) between 5 to 10 %   (*d*) zero.

45. The operating speed of a synchronous motor can be changed to a new fixed value by
    (*a*) changing the load   (*b*) changing the supply voltage
    (*c*) changing the frequency   (*c*) using brakes.

46. In case of turbo alternator the rotor is usually made of
    (a) cast iron   (b) forged steel
    (c) laminated stainless steel   (d) High speed steel.

47. The span for a full pitch coil wound for six poles is
    (a) 180deg. mech.   (b) 90 deg. mech.
    (c) 60 deg. mech.   (d) 45 deg. mech.

48. The pitch factor for a two third short pitch coil is
    (a) 0.5   (b) 0.66   (c) 0.866   (d) 0.707

49. The armature flux helps the main field flux when the load power factor is
    (a) unity   (b) zero-lagging   (c) 0.8 leading   (d) zero leading.

50. A commercial alternator has
    (a) rotating armature and stationary field
    (b) stationary armature and rotating field
    (c) both armature and field roating
    (d) both armature and field stationary.

51. pitch factor for 5/6 short pitch coil is
    (a) 0.966   (b) 0.833   (c) 1.0   (d) 0.75.

52. Distribution factor for a winding having 3 slots/pole/phase and a slot angle of 20 degree is
    (a) 0.96   (b) 1.0   (c) 0.5   (d) 0.707.

53. The speed regulation of synchronous motor is
    (a) unity   (b) zero   (c) less than 5%   (d) infinite.

54. Synchronous motor is used for
    (a) constant load   (b) constant speed
    (c) high starting torque   (d) variable speed.

III. Fill in the Blanks :

1. The stator winding of a three phase alternator is always .... connected.

2. The rotor of an alternator has .... slip rings for .... supply.

3. A synchronous motor can run .... of power factor.

4. In a synchronous motor the magnitude of E_b can be varied by increasing ....

5. A synchronous motor working on leading power factor and not coupled to any mechanical load is called ....

6. Hunting of a synchronous motor usually results from ....

7. Synchronous reactance is the sum of .... reactance and .... reactance.

8. For an salient pole alternator X_d is .... X_q.

9. Ratio of X_d/X_q for cylindrical rotor machine is .... X_d/X_q of salient pole machine.

10. X_d and X_q are determined by .... test.

11. In synchronous machine the value steady state reactance is .... transient and sub transient reactance.

12. The armature voltage of a three phase alternator supplying power consisting of .... sequence component.

13. Zero sequence component current produces .... magnetic field in the stator.

14. Sub transient reactance is due to the presence of ....

15. .... type of rotor is best suited for turbo alternator because ....

16. At .... power factor the armature current of synchronous motor is minimum.

17. If the calculated value of Z_s is higher than actual value then voltage regulation of synchronous machine will be .... as compared to the actual value.

18. For constant kVA, lagging power factor requires .... field current than leading power factor I_a.

19. Turbo alternator usually have .... pole.

20. The D/L ratio of the rotor of turbo alternator is .... than hydro alternator.

Answers

II. Multiple Choice Questions :

1. (c)	**2.** (a)	**3.** (d)	**4.** (d)	**5.** (d)
6. (a)	**7.** (b)	**8.** (c)	**9.** (d)	**10.** (b)
11. (a)	**12.** (a)	**13.** (b)	**14.** (b)	**15.** (c)
16. (c)	**17.** (a)	**18.** (d)	**19.** (a)	**20.** (c)
21. (a)	**22.** (b)	**23.** (b)	**24.** (c)	**25.** (a)
26. (a)	**27.** (b)	**28.** (d)	**29.** (c)	**30.** (d)
31. (d)	**32.** (d)	**33.** (a)	**34.** (b)	**35.** (a)
36. (b)	**37.** (a)	**38.** (c)	**39.** (a)	**40.** (c)
41. (d)	**42.** (c)	**43.** (a)	**44.** (d)	**45.** (c)
46. (b)	**47.** (c)	**48.** (c)	**49.** (d)	**50.** (b)
51. (a)	**52.** (d)	**53.** (b)	**54.** (b)	

SECTION–D

2.7   SPECIAL DRIVES

In addition to classic electrical machine such as DC machine, induction or synchronous machine, there are some new types of electric machines created in the last few years. These machines of low weight and small in size and suitable for special drives. The following machines belong to this category stepper motor, switched reluctance motor, single phase induction motor, and modular permanent magnet machine etc.

2.7.1   Stepper Motor

This special type of synchronous machine is mainly used as positioning drives for all kinds of controls. This is also used in printers and typewriters in many different ways. The digital control of the stator winding leads to a rotation of the rotor shaft about the step angle *a* for each current pulse, so that for *n* control instructions the total angle *n* ×*a* is covered at the shaft. Stepper motors enable positioning without feedback of the rotor position, which cannot be achieved using DC servo drives or three-phase servo drives.

The basic configuration and the method of operation of stepper motors is shown in Figure 2.103. A permanent-magnet rotor is arranged between the poles of two independent stator parts. Both of the stator parts consist of a winding with centre tap that means two halves of the winding. Any part can be supplied with current by the transistors T_1 to T_4. If, for example, the transistor T_1 is switched on, there is a north pole on the top of stator 1 and a south pole at the bottom. If transistor T_3 is switched on at the same time north pole is on the right and south pole on the left side of stator 2. That means the rotor turns to the position shown in the Figure 2.103.

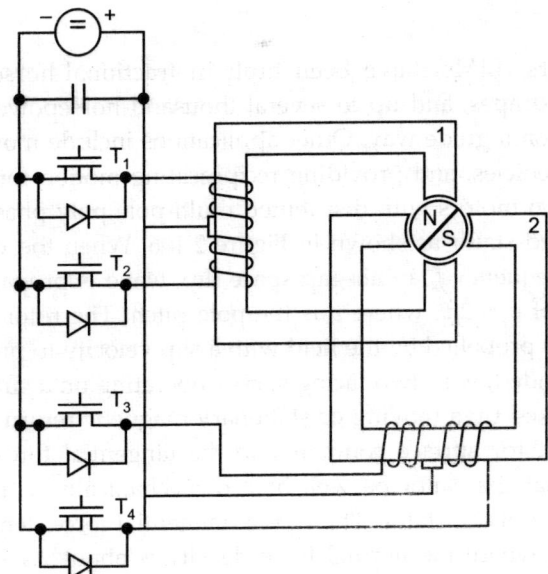

FIGURE 2.103. *Stepper motor*

The primary characteristic of a stepper motor is its ability to rotate a prescribed small angle (step) in response to each control pulse applied to its windings. Below about 200 pulses per second, the motor rotates in discrete steps in synchrony with the pulses; at higher frequencies up to 16,000 pulses per second, the motor slews without stopping between pulses. Although motors are available for step angles of 90° to 180°, the common step is 1.8°. Stepper motors are categorized as permanent magnet rotor (PM), variable reluctance (VR), or hybrid (PM-VR). The rotor of the PM aligns itself with the energized stator poles as shown in Figure 2.104(b). The rotor turns until the poles are aligned at each step. The PM-VR hybrid shown in Figure 2.104(c) has a high skew rate yet retains holding torque when the power is turned off. Motors can be made to rotate in half-steps to increase accuracy. Performance of stepper motors is described by two types of curves: the pull-out torque vs. speed curve, as shown in Figure 2.105.

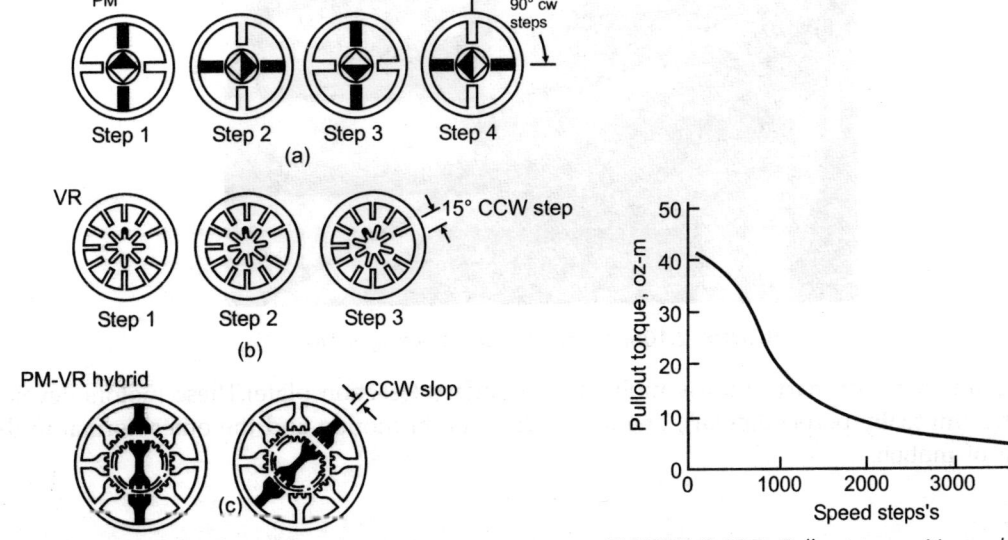

FIGURE 2.104. *Working of Stepper motor*

FIGURE 2.105. *Pull out torque Vs speed steps*

2.7.2 Linear Motors

Linear induction motors (LIMs) have been built in fractional-horsepower ratings for such applications as moving drapes, and up to several thousand horsepower for driving tracked air-cushion transit vehicles on a guide way. Other applications include moving freight cars in yards, driving people-mover vehicles, and providing reciprocating motion for machine tools. LIMs are built like rotary induction motors with distributed multi-pole poly phase windings placed in the slots of a plane laminated stator as shown in Figure 2.106. When the windings are excited by a poly phase voltage of frequency f, an air-gap space flux wave is propagated along the length of the stator at a velocity of $v = 2fp$, where p is the pole pitch. The rotor consists of an aluminum or copper sheet, which is propelled by the field with a slip velocity to provide the required thrust. LIMs are either double-sided, with two facing stators operating on a single rotor, or single-sided, with the rotor sheet backed by a moving or stationary magnetic return path. The magnetic force density normal to the stator surface compared to the tangential force density that moves the rotor, which requires that the stator be well braced mechanically to maintain constant air-gap distances over the surface of the stator. The typical tangential force density is about 0.24 kg/cm^2 for air-cooled windings, where the normal force density is about 2.4 kg/cm^2.

Rotary, squirrel cage induction motor, split radially along its axis of rotation and flatten out is a linear induction motor that produces direct linear force instead of torque. Linear Induction Motor is a non-contacting, high speed, linear motor that operates on the same principle as a rotary squirrel cage induction motor. It is capable of producing speeds up to 45 m/s and, therefore, useful in applications where accurate positioning is not required.

FIGURE 2.106. *Construction and Working of LIM*

Diagram shows primary coil assembly and secondary (reaction plate).These motors develop two forces (mutually perpendicular), one in the direction of motion and the other normal to the direction of motion.

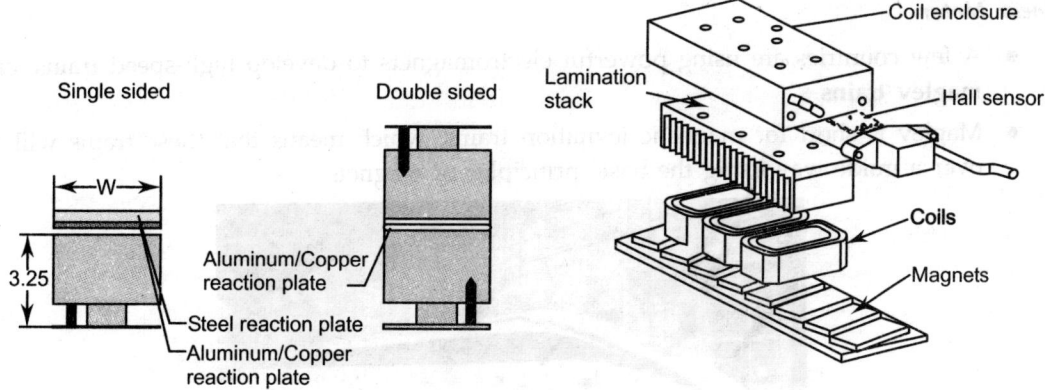

FIGURE 2.107

The normal force may be an attraction or a repulsion force between the primary and secondary. A machine in which the net force is such that the secondary tends to be suspended over the primary may be used mainly for suspension and called a linear levitation machine (LLM). Conversely, a machine used primarily for producing thrust is called a linear motor Both LIM's and LSM's may be used a levitation or as linear motors.

Advantages :

1. Size is small, compact and therefore it fits into small spaces.
2. No backlash from gears or slippage from belts – provides smooth operation.
3. Reliability *i.e.*, non-contact operation reduces component wear and maintenance of machine.
4. Wide speed range.
5. Design compatibility with either a moving coil or moving magnets.
6. Ease of Control and Installation.

Limitations :

- Costly to purchase and installs
- **Force Per Package Size :** Linear motors are not compact force generators compared to a rotary motor.
- **Heating :** The force is often attached to the load. If an application is sensitive to heat, thermal management techniques need to be applied.

Applications :

- Linear applications (lower precision)
- Conveying Systems
- Cranes Drives
- Baggage Handling
- Personal Rapid Transport Systems
- Theme Park Rides

Linear Motor

- A few countries are using powerful electromagnets to develop high-speed trains, called **maglev trains**.

- Maglev is short for magnetic levitation trains, which means that these trains will float over a guide way using the basic principles of magnets.

FIGURE 2.108

The principle of a Magnet train is that floats on a magnetic field and is propelled by a linear induction motor. A maglev train floats about 10mm above the guide way on a magnetic field. It is propelled by the guide way itself rather than an onboard engine by changing magnetic fields.

2.7.3 Hysterisis Motors

By constructing the secondary core of an induction motor of hardened magnet steel, in place of the usual annealed low-loss silicon-steel laminations, the secondary hysterisis can be greatly magnified, producing effective synchronous motor action. Such hysterisis motors, having smooth rotor surfaces without secondary teeth or windings, give extremely uniform torque, are practically noiseless, and give substantially the same torque from standstill all the way up to synchronous speed. A hysterisis motor is a true synchronous motor with its load torque produced by an angular shift between the axis of rotating primary mmf and the axis of secondary magnetization. When the load torque exceeds the maximum hysterisis torque, the secondary magnetization axis slips on the rotor, giving the same effect as a friction brake set for a fixed torque.

Despite the interesting characteristics of this type of motor, it is limited to small sizes, because of the inherently small torque derivable from hysterisis losses. Only moderate flux densities are practicable, owing to the excessive excitation losses required to produce high densities in hard magnet steel, and, therefore, about 40 W/Kg of rotor magnet steel represents the maximum useful synchronous power on 50 Hz. Hysterisis motors have found an important use for phonograph-motor drives, their synchronous speed enabling a governor to be dispensed with and freedom from tone waver to be secured. The Telechron motor, which is so widely used for operating electric locks, also operates on the hysterisis-motor principle. In the Telechron motor, a two-pole rotating field is produced in a cylindrical air space, and into this space is introduced a sealed thin-metal cylinder containing a shaft carrying one or more hardened magnet-steel disks, driving a gear train. The magnetic field causes the steel disks to revolve at 3000 rpm, driving through the gears a low-speed shaft, usually 1 rpm, which merges from the sealed cylinder through a closely fitting bushing designed to minimize oil leakage. Although the magnetic field has to cross a very considerable air-gap length and pass through the thin walls of the metal cylinder, the power required to drive a well-designed clock is so small that sample output is obtained with only about 2 W input for ordinary household-clock sizes. The hysterisis motor has been displaced for phonograph and tape reel drives by the transistor-driven brushless dc motor. It has been displaced for electric clocks by solid-state circuits with digital readout.

2.7.4 Universal Motors

Small series motors up to about half HP rating are commonly designed to operate on either direct current or alternating current and so are called universal motors. Universal motors may be either compensated or uncompensated, the later type being used for the higher speeds and smaller ratings only. Owing to the reactance voltage drop, which is present on alternating current but absent on direct current, the motor speed is somewhat lower for the same load ac operation, especially at high loads. On alternating current, however, the increased saturation of the field magnetic circuit at the crest of the sine wave of current may materially reduce the flux below the dc value, and this tends to raise the ac speed. It is possible, therefore, to design small universal motors to have approximately the same speed-torque performance over the operating range, for all frequencies from 0 to 50 Hz. On a typical compensated-type quarter HP motor, rated at 3000 rpm, the 50-Hz speed may be within 2% of the dc speed at full-load torque but 15% or more lower at twice normal torque, while on an uncompensated motor the speed drop will be materially greater.

The commutation on alternating current is much poorer than on direct current, owing to the current induced in the short-circuited armature coils, and this provides a definite limitation on their size and usefulness. If wide brushes are used, the short-circuited currents are excessive and the motor-starting torque is reduced, while if narrow brushes are used, there may be excessive brush chatter at high speeds, causing short brush life. Good design, therefore, requires careful proportioning of commutator and brush rigging to meet conflicting electrical, mechanical, and thermal requirements. Universal motors are generally used for vacuum cleaners, portable tools, food mixers, and similar small devices operating at maximum speeds of 3000 to 10,000 rpm.

2.7.5 Review Exercises

I. Answer in Brief :
1. Name the electrical factors that govern the selection of the motors.
2. Choose the suitable motors for the following jobs -
   (*a*) Lathe   (*b*) Hoist
   (*c*) Winding the paper reel in paper mill   (*d*) Electrical Watches
   (*e*) Electric trains   (*f*) Prime mover of D.C. Generator.

II. Multiple Choice Questions :
1. In hand tool applications the single phase motor used is
   (*a*) shaded pole motor   (*b*) capacitor start motor
   (*c*) capacitor run motor   (*d*) AC series motor
2. In a particular application needs high speed and high starting torque, which type of motor will be preferred
   (*a*) universal motor   (*b*) capacitor start motor
   (*c*) shaded pole motor   (*d*) split phase motor
3. Which type of motor has the highest power to weight ratio
   (*a*) universal motor   (*b*) capacitor start motor
   (*c*) capacitor run motor   (*d*) none.

4. The machinery in which starting torque as compared to rated torque is maximum is
   (a) Fans   (b) Blowers   (c) Loaded crusher   (d) Printing presses
   (e) Machine tools.

5. In which case DC shunt motor will be suitable
   (a) Coal cutting machine   (b) Traction
   (c) Centrifugal pump   (d) Machine tools.

6. Cumulative compound DC motors are suitable for except
   (a) Cool cutting machine   (b) Hoist and cranes
   (c) Centre lathes   (d) Traction.

7. Which of the following motors has high starting torque
   (a) DC shunt motor   (b) DC series motor
   (c) Induction motor   (d) Synchronous motor

8. Which of the following motor will be suitable for small home air conditioners
   (a) Hysterisis motor   (b) Two value capacitor
   (b) Shaded pole motor   (d) universal motor.

9. The speed of universal motor is commonly reduced by using
   (a) Brakes   (b) Chains   (c) Belts   (d) Gearing.

Answers

II. Multiple Choice Questions

1. (d)   2. (a)   3. (a)   4. (c)   5. (d)
6. (a)   7. (b)   8. (b)   9. (c)

SECTION–E

2.8   THREE PHASE INDUCTION MOTOR

2.8.1   Introduction

The induction motor was invented by Nikola Tesla in 1882 in France. The induction motor with a cage was invented by Mikhail Dolivo-Dobrovolsky about a year later in Europe. Currently, the most common induction motor is the cage rotor motor.

FIGURE 2.109. *Induction motors*

2.8.2   Classification

1. Based on type of supply
    (*i*) Three phase induction motor (self starting by nature)
    (*ii*) Single phase induction motor (not self starting)
2. Based on rotor construction
    (*i*) Squirrel-cage rotor induction motor
    (*ii*) Slip ring wound rotor induction motor

The most common rotor is a squirrel-cage rotor. It is made up of bars of either solid copper (most common) or aluminum that span the length of the rotor, and are connected through a ring at each end. The rotor bars in squirrel-cage induction motors are not straight, but have some skew to reduce noise and harmonics.

2.8.3   Principle of Operation of Three Phase Induction Motor

A 3-phase supply is given to the distributed stator windings of induction motor, which are 120° apart, produces a rotating magnetic field in an induction motor. When this rotating magnetic flux links the stationary rotor bars (conductors), emf induced due to relative motion between them. The rotor bars are short circuited in squirrel cage induction motor. Hence, the rotor current starts flowing and rotor flux is produced. This rotor flux opposes the cause by which it is produced (Lenz's Law). Hence, rotor starts rotating. When the rotor speed reaches near synchronous speed the rotor current diminishes and correspondingly the torque is also reduced. Therefore, induction motors can not operate at synchronous speed and called asynchronous motors. This difference between the speed of the rotor and speed of the rotating magnetic field in the stator is called *slip*. Slip is a unitless quanity and is the ratio between the relative speeds of the magnetic field as seen by the rotor to the speed of the rotating field.

FIGURE 2.110. *Cut section of induction Motor*

The synchronous speed (speed of rotating field), n_s, is given by:

$$n_s = \frac{120\,f}{P}$$

where, f – the supply frequency in Hz,

P – the number of pole pairs.

The slip speed is the difference between synchronous speed and the rotor speed.

$$\text{Slip speed} = (n_s - n) \text{ rpm, where } n \text{ is the rotor speed.}$$

$$\text{Percentage } slip \ s = \frac{\text{Slip speed}}{\text{Synchronous speed}} = \frac{n_s - n}{n_s} * 100$$

The rotational speed of the rotor is controlled by the number of pole pairs (number of windings in the stator) and by the frequency of the supply voltage. Before the development of cheap power electronics, it was difficult to vary the frequency to the motor and therefore the use of variable speed induction motor were limited.

There are various techniques to produce a desired speed. The most commonly used technique is PWM (Pulse Width Modulation), in which a DC signal is switched on and off very rapidly, producing a sequence of electrical pulses to the inductor windings. The duty cycle of the pulses determines the average power input to the motor. For example, a 100 V DC signal that is cut into on- and off- pulses of equal width, has an average voltage of 50 V. If the on- pulses are one third of the duration of the off pulses, the average would be 25 V. The frequency of the pulses determines the motor speed.

Besides the simplicity and ruggedness of the induction motor, it has no brushes and is easy to control, many older DC motors are being replaced with induction motors and accompanying PWM inverters in industrial applications.

2.8.4 Starting of Induction Motor

At the time of starting of the induction motor, the slip is equal to 1 as the rotor speed is zero. Hence, there is maximum relative motion between the rotor conductor and rotating magnetic flux and the induced emf in the rotor is maximum. As a result, a very high current flows through the rotor bars. This is similar to a transformer with the secondary coil short circuited, which causes the primary coil to draw a high current from the mains. The starting current is of the order of 5 to 10 times the full load current. This high starting current can damage the motor windings and also it causes heavy drop in the line voltage, which affects the performance of other appliances connected to the same line. To avoid such effects, the starting current should be limited using starters. Starter limits the starting current by providing reduced voltage to the motor. As the rotor speed increases, the full rated voltage is applied to it.

FIGURE 2.111. *Induction motor with cooling fins*

Types of starters

1. Direct on line starter
2. Auto-transformer starter
3. Star-Delta starter
4. Stator resistance/reactance starter

These methods generally not used to start the slip ring induction motors. In slip ring induction motor, the rotor windings are connected to slip rings, which make it different from the squirrel-cage rotor. The slip ring induction motors the rotor resistance is varied to control the strating current, which also produce high starting torque at the time of starting.

2.8.5 Working Principle of Slip Ring Induction Motor

The working principle of slip ring induction motor is same as squirrel-cage induction motor. In these motor, rotor winding is distributed just like the stator winding. The rotor has large number of turns, so that induced voltage is higher for the same field strength compared to squirrel cage. The rotor windings are connected in star and the three points of star connected to three slip rings. At the time of starting these slip rings are short circuited through star connected variable resistors to limit the inrush current and gradually the resistance value is reduced as speed increases as back emf build up. At full speed the slip rings are short circuited together.

The main advantage of slip ring induction motor is its higher starting torque and lower starting current. Today slip ring motor is mostly superseded by induction motors with variable-frequency drive where wide smooth speed variation is required.

The four quadrantal torque – slip characteristic of induction machine for different modes of operation is shown Figure 2.112.

The torque Equation for 3-phase induction motor is

$$T = \frac{ks\,E^2\,R_2}{\left(R_2^2 + (sX_2)^2\right)}$$

At the time of starting slip is unity and $R_2 << sX_2$ therefore $T_{start} = \dfrac{kE^2\,R_2}{sX_2^2}\,\alpha\,\dfrac{1}{s}$.

Hence, the starting torque is inversely proportional to slip.

On the other hand, under running conditions the speed is close to synchronous speed and

the slip $s = 1 - 5\%$. Therefore, $R_2 >> sX_2$ and $T_{start} = \dfrac{ks\,E^2}{R_2^2}\,\alpha \dots s$.

Hence, the torque at the time of running is directly proportional to the slip.

The maximum torque may be obtained when rotor resistance is equal to rotor reactance at a particular slip *i.e.*, $R_2 = sX_2$.

Similarly maximum efficiency may be obtained when variable losses (copper losses) equals to constant losses (iron losses, friction and widage losses) at a constant speed.

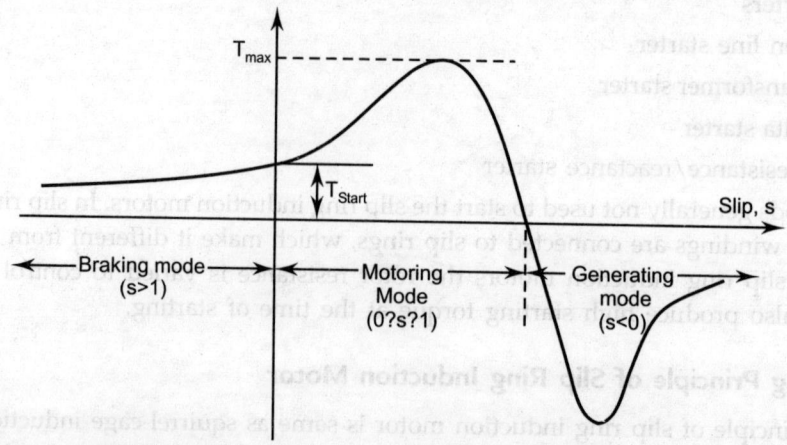

FIGURE 2.112. *Torque Slip Characteristics of Induction motor*

2.8.6   Effect of Rotor Resistance

In case of squirrel cage induction motor the rotor resistance is assumed constant during its operation, although there is a slight variation due to change in temperature.

In slip ring induction motor, the rotor resistance may be changed during its operation by inserting an external resistance through slip rings. Hence, different starting torque may be obtained by suitably adjusting the value of rotor resistance. The speed-torque characteristics of induction motor for different values of rotor resistance are shown in Figure 2.113.

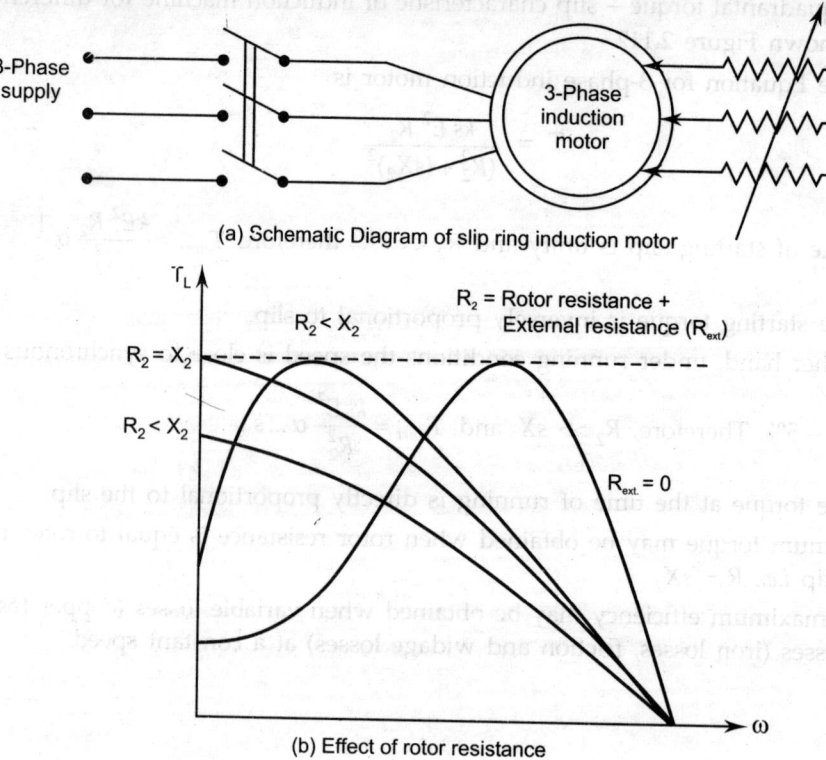

(a) Schematic Diagram of slip ring induction motor

(b) Effect of rotor resistance

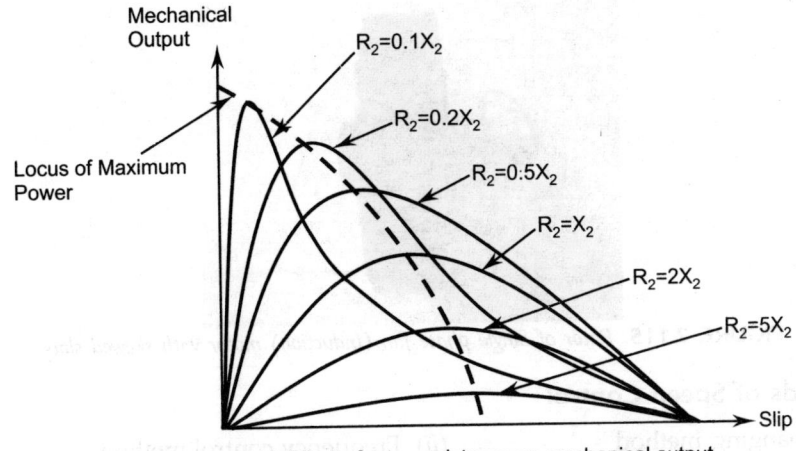

(c) Effect of rotor resistance on mechanical output

FIGURE 2.113. *Effect of rotor resistance on mechanical output and torque*

The mechanical output = $T^*\omega$ is zero at synchronous speed at which torque produced by induction motor $T = 0$ and also at starting of induction motor when speed $\omega = 0$ *i.e.,*

At slip $s = 1$ means starting mechanical power output = 0 and at slip $s = 0$ means synchronous speed again mechanical power output = 0 as shown in Figure 2.113 (c).

2.8.7 Crawling and Cogging Phenomenon

The plain squirrel cage motor may exhibit peculiar behavior in starting especially when stator slots S_1 equals to the integral multiple of rotor slots S_2. In this situation the reluctance which is a function of space produce an alignment forces (torque) more then starting forces (torque). Due to these torques, motor unable to start. This phenomenon is called *cogging*. This problem can be overcome by taking suitable combination of stator and rotor slots at the time of design; and using skewed slots as shown in Figure 2.115.

The harmonics in power supply also affect the starting of induction motor, specially fifth and seventh harmonics. The fifth harmonic flux rotates in backward direction with one fifth of synchronous speed. Seventh harmonic flux rotates in forward direction with one seventh of synchronous speed. These harmonic MMFs superimposed on the fundamental MMF and the saddle effect is observed in torque slip characteristic at $1/7^{th}$ of synchronous speed. If the load torque characteristic intersect the motor characteristic, motor shows stable operation at this speed, which is much lower than its full speed as shown in Figure 2.114. This phenomenon is called *crawling*.

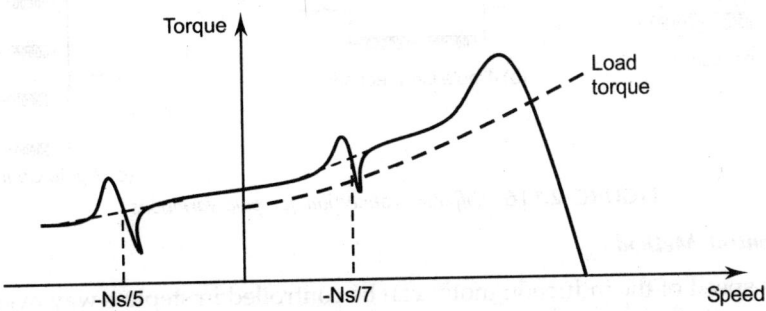

FIGURE 2.114. *Effect of harmonics on speed torque characteristic*

FIGURE 2.115. *Rotor of Single phase fan (Induction) motor with skewed slots*

2.8.8   Methods of Speed Control

(*i*) Pole changing method   (*ii*) Frequency control method

(*iii*) Varying applied voltage   (*iv*) Cascade connection

(*v*) Rotor rheostat control   (*vi*) Injecting EMF in rotor circuit.

First three methods used for both squirrel cage as well as slip induction motors, while last two methods only applicable to slip ring induction motors. In cascade connection, at least one slip ring induction motor is necessary.

(*i*) Pole Changing Method

In this method separate stator winding connections is required for each speed. The speed variation in this method is always in the ration of 2:1, because number of poles could change in multiples of two.

When the number of poles reduces by connecting the windings in parallel, the speed increased. Similarly, the number of poles increases by connecting the windings in series to reduce the speed. In this method the rotor cage resistance also changes due to the change in end ring resistance. The speed variation is only in steps.

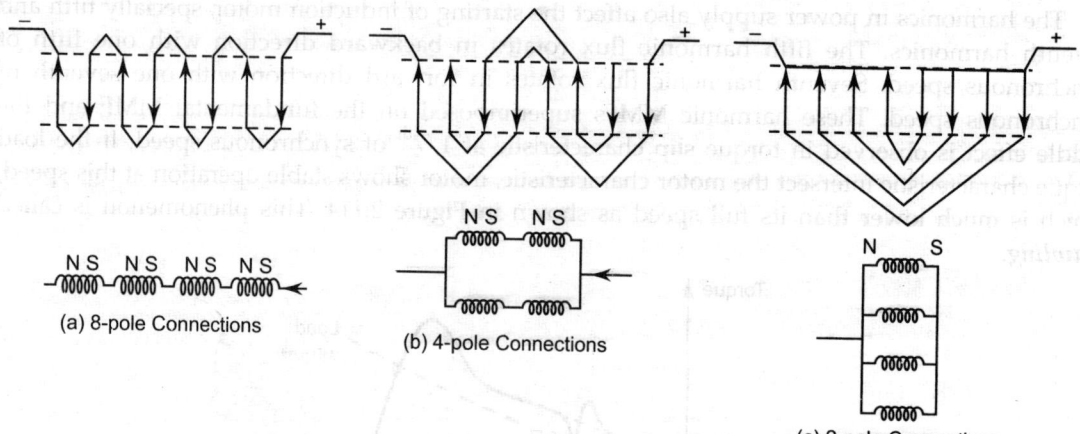

(a) 8-pole Connections

(b) 4-pole Connections

(c) 2-pole Connections

FIGURE 2.116. *Different connection for pole variations*

(*ii*) Frequency Control Method

The synchronous speed of the induction motor can be controlled in stepless way over a wide range by changing the supply frequency. The frequency may be controlled by either controlling the speed of generator or with the help of solid state devices. The frequency control using solid state devices is economical option. Earlier on warships, where cruishing and fighting speeds are required, a low frequency generator is provided for low speeds and separate and more powerful generators of higher frequencies for high speeds. **The frequency is reduced to reduce the speed of induction motor without controlling the terminal voltage, then the maximum torque also resuces, due to reduction in air gap flux as shown in Figure 2.118.**

The resultant air gap flux per pole may be written as

$$\phi_r = \frac{1}{4.44 \, K_w \, N_{ph}} \left(\frac{V}{f} \right)$$

To keep the flux constant, (V/f) ratio must be kept constant, so that the maximum torque should remain constant as shown in Figure 2.117.

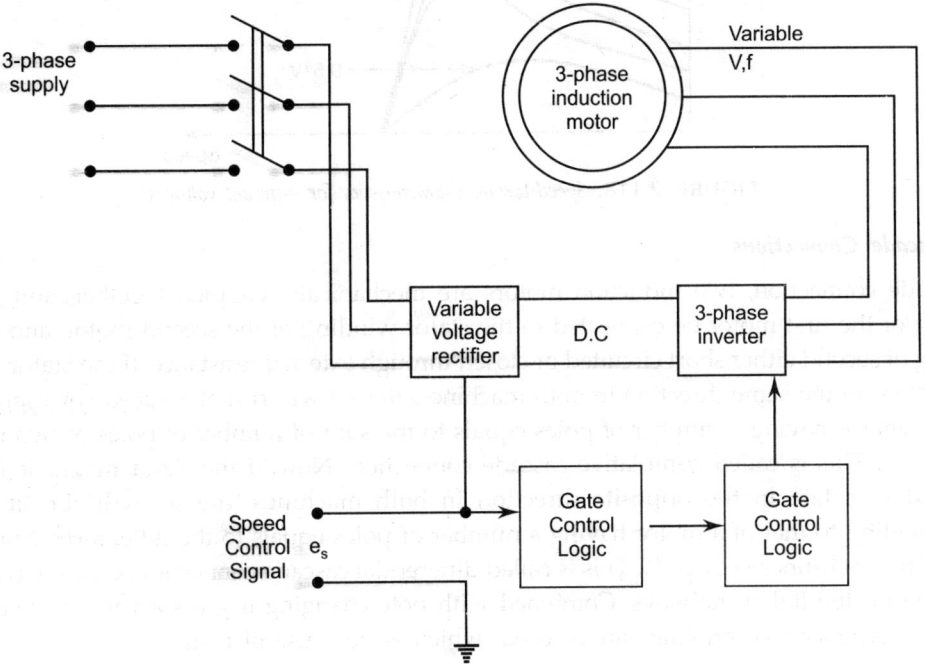

(a) Schematic diagram of frequency control method

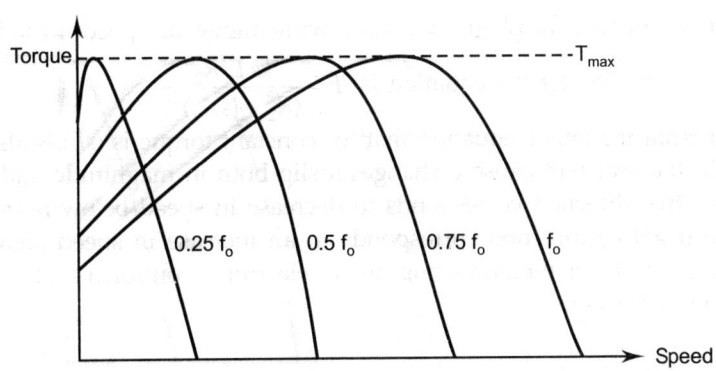

(b) T-ω characteristics for variable frequency for constant V/f

FIGURE 2.117

(iii) Changing Applied Voltage

From the torque equation $T = \dfrac{ks\,E^2R_2}{(R_2^2 + (sX_2)^2)}$ the torque is directly proportional to the square of the applied voltage $T \propto E^2 \propto V^2$. Hence, the torque is very sensitive to the supply voltage. A change of 5% in supply voltage will produce a change of approximately of 10% in rotor torque.

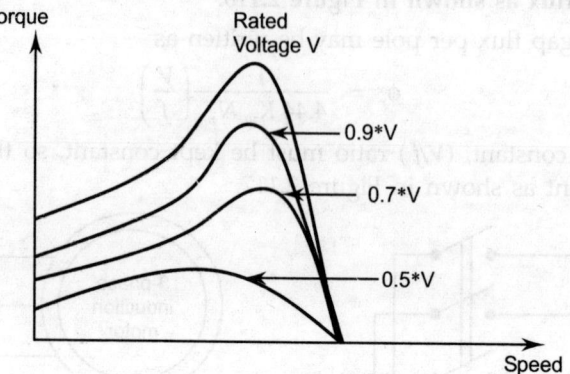

FIGURE 2.118. *Speed-torque characteristics for different voltages*

(iv) Cascade Connections

In cascade connection, two induction motors are mechanically coupled together, and the rotor winding of the first motor be connected to the stator winding of the second motor, and the rotor winding of second either short circuited or closed through external resistance. If the Stator magnetic fields rotate in the same direction in both machines, the set will run at a speed corresponding to that of a motor having a number of poles equals to the sum of number of poles of two machines i.e., $P_1 + P_2$. This is called cumulative cascade connection. Now, if the stator magnetic fields are arranged to rotate in the opposite direction in both machines, the set will run at a speed corresponding to that of a motor having a number of poles equals to the difference of number of poles of two machines i.e., $P_1 - P_2$. This is called differential cascade connection. Cascade connection was used on the Italian railways. Combined with pole changing it gives a variety of speed and of course, regenerative braking can be used, which is very useful feature.

(v) Injecting EMF in the Rotor Circuit

This method of speed control gives a very wide range of speed variation below and above synchronous speed. The torque equation is $T = \dfrac{ks\,E^2R_2}{(R_2^2 + (sX_2)^2)}$.

It is clear from the above equation that for constant torque is (sE) is also constant. Therefore, any change in the emf will cause a change in slip both in magnitude and direction. Increase in slip in the positive direction corresponds to decrease in speed below normal speed an decrease in slip in the negative direction corresponds to an increase in speed above normal speed. The schematic diagram of this method using power electronic controllers called static Kramer system is shown in Figure 2.119.

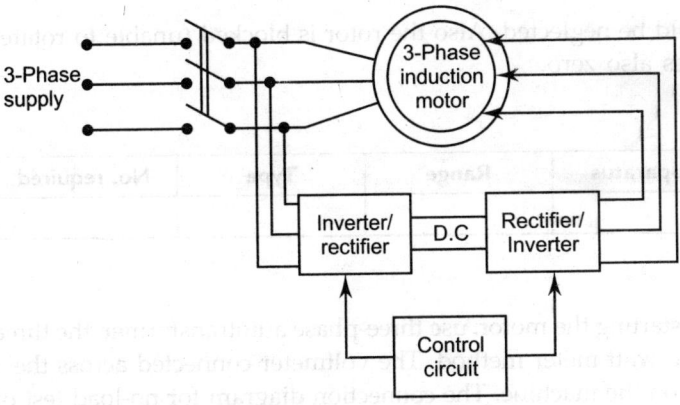

FIGURE 2.119. *Schematic diagram of static Kramer System*

2.8.9 Performance Characteristics of Induction Motor

There are certain performance indices in the induction motor *i.e.,* variation in speed from no load to full load (slip), power factor variation, efficiency etc.

At no-load

$$\text{Slip H} \approx 0$$
$$\text{Power Factor} = 0.2 - 0.3 \text{ (very low)}$$
$$\text{Efficiency} = 0 \text{ (because the output is zero)}$$

At full load

$$\text{Slip} = 3 - 5\%$$
$$\text{Power Factor} = 0.8 - 0.85$$
$$\text{Efficiency} = 80\% \text{ (the maximum efficiency at nearly 85\% of full load)}$$

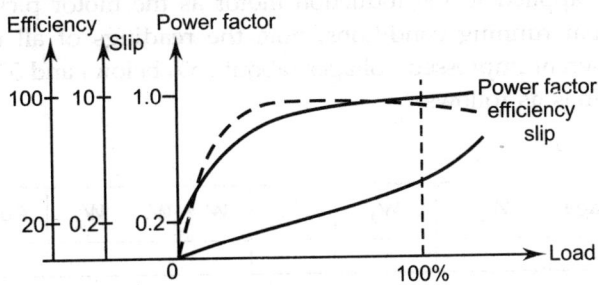

FIGURE 2.120. *Performance characteristics of induction motor*

2.9 EXPERIMENT ON INDUCTION MOTORS

2.9.1 To Perform No-load and Locked Rotor Tests on Induction Motor

Theory

The power input during the no-load test represents iron losses and friction & windage losses in the machine for the particular applied voltage (Assuming that the copper losses are negligibly small. Similarly, the power input during the blocked rotor test represents copper losses for the particular value of the current, because in blocked rotor test the applied voltage is very small and the iron losses could be neglected. Also the rotor is blocked (unable to rotate), hence, the friction and windage losses also zero.

Apparatus

S. No.	Apparatus	Range	Type	No. required	Make

Connections

For the purpose of starting the motor, use three-phase autotransformer, the three-phase power input is measured by two watt meter method. The voltmeter connected across the supply indicates the impressed voltage on the machine. The connection diagram for no-load test of induction motor is shown in Figure 2.121.

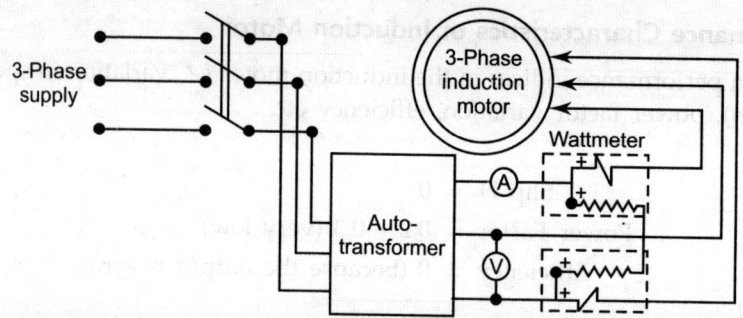

FIGURE 2.121. *Conection diagram for No-load Test of Induction Motor*

Procedure

NO-LOAD TEST : To start the squirell cage induction motor using auto-transformer by gradually increasing the voltage applied to the induction motor as the motor picks up speed. When the motor comes to normal running conditions, note the readings of all the meters. Repeat the experiment for the different impressed voltages (about 25% below) and 5% above the rated value and tabulate the readings as follows :

Observations

S,No.	Applied voltage	W_1	W_2	I	$W = W_1 + W_2$	Power factor = $W/\sqrt{3}EI$

BLOCKED ROTOR TEST : Following steps are to be followed during this test :

  (*i*) Hold the rotor preventing it from rotation.

  (*ii*) Apply small voltage to induction motor (about 2 to 3% of the rated value).

  (*iii*) Note down the readings of different meters.

  (*iv*) Increase the supply voltage in steps till the full load current flows through the machine.

  (*v*) Note the readings of the various meters for different impressed voltages.

Observations

S.No.	Applied voltage	W_1	W_2	I (stator current)	$W = W_1 + W_2$	Power factor = $W/\sqrt{3}EI$

Graphs

Following graphs are required to be drawn from the above results :

(*i*) Stator current versus applied voltage.

(*ii*) Stator current versus input (watts).

(*iii*) Stator current versus power factor.

EXPERIMENT NO. 18

2.9.2 To Perform Load Test on Induction Motor (Squirrel Cage or Slip Ring Type)

The aim of the experiment is to determine the (*i*) efficiency, (*ii*) slip, (*iii*) power factor, (*iv*) current of the motor at different outputs (kW) by actually loading the motor.

Connections

The induction motor is directly coupled to a DC generator for the purpose of loading the motor. Three-phase power input to the motor is measured by two-wattmeter method using suitable current transformer. The line currents are indicated by the ammeters connected to the secondary of current transformer. DC generator is loaded by using trolly load and output is measured by an ammeter and voltmeter. Slip frequency is then determined with the help of speed measured from tachometer.

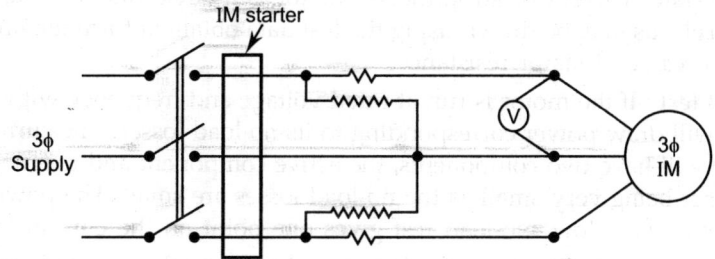

FIGURE 2.122. *Connection diagram for IM*

Procedure

Following steps are to be followed during this test :

(*i*) Start the induction motor using auto-transformer or star-delta starter keeping the generator field switch open.

(*ii*) Note down the readings of AC ammeters and voltmeters. Wattmeters readings indicate the power input to the motor running under no-load.

(*iii*) Determine slip with the help of tachometer.

(*iv*) Close the field switch of the generator and excite it to its rated voltage and load the generator by putting the load on.

(*v*) Keep the generator terminal voltage constant for all loads. Also try to maintain the AC impressed voltage to the induction motor constant.

(*vi*) Note down the readings of instruments for various loads and calculate the slip each time.

Graphs

Following graphs are required to be drawn from the above results :

    (*i*) Output (kW) versus power factor.

    (*ii*) Output (kW) versus % slip.

    (*iii*) Output (kW) versus current.

    (*iv*) Output (kW) versus efficiency.

Observations

S. No.	V(A.C.)	W_1	W_2	Input (W)	Speed	I_L	V(D.C.)	Efficiency (η)

The wattmeter connected to AC supply indicates the input to the motor. Calculate the input to the generator at various loads by knowing the efficiency of generator (assuming overall efficiency 87%). This input must be output of the induction motor. As the input and output of motor are known, efficiency of the motor can be calculated for various loads.

EXPERIMENT NO. 19

2.9.3 Circle Diagram of Induction Motor from No-load and Blocked Rotor Tests

Theory

The locus of the stator current of an induction motor is a circle under certain reasonably valid assumptions. This locus may be drawn using the test data obtained form the no-load and blocked-rotor test and the value of stator resistance.

    (*a*) **No-load test :** If the motor is run at rated voltage and frequency without any mechanical load, it will draw power corresponding to its no-load losses. The current drawn (no-load current) will have two components, the active component and magnetizing components, the former being very small as the no-load losses are small. The power factor at no load is, therefore very low. No-load test gives one point on the current locus.

    (*b*) **Blocked-rotor test :** This test affords a second point on the current locus, and is analogous to the short-circuit test on a transformer. The stator is supplied with a low voltage of rated frequency, and with the rotor blocked (*i.e.*, physically prevented from moving) the power input, current input and voltage applied are recorded. The data when converted to rated voltage gives the short-circuit current and the power factor.

The power input during blocked-rotor test is almost wholly consumed in the stator and rotor copper losses. From the short-circuit current and the power input, the total equivalent resistance of the stator and rotor can be obtained.

From this if the stator resistance (which may actually be measured) is subtracted, the remainder is the rotor resistance referred to stator.

If R is the d.c. resistance of the stator, it's a.c. resistance or effective resistance will usually be higher on account of the skin effect. Usual range of effective resistance is 1.05 to 1.2 times d.c. resistance, the exact value depending upon the frequency and conductor size.

From the values of the no-load and blocked-rotor currents at rated voltage, the corresponding power factors and the equivalent stator and rotor resistance, the complete circle diagram may be drawn as follows :

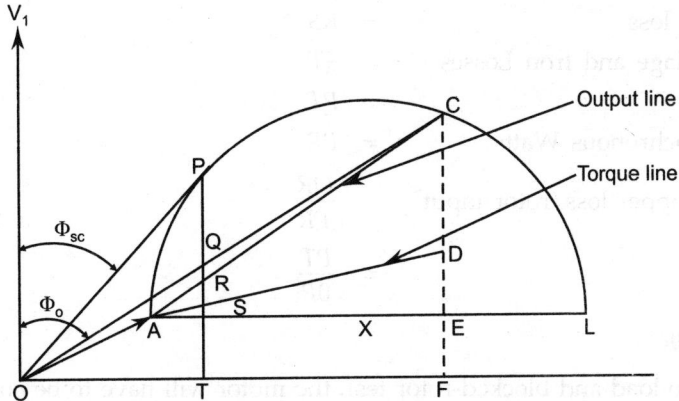

FIGURE 2.123. *Circle diagram of an Induction motor.*

With V_1 as reference, OA is plotted equal to no-load current I_0 at an angle ϕ_0 lagging, where $\cos \phi_0 = \dfrac{P}{\sqrt{3}VI}$, and P_0 is the no-load power input at rated vulgate and frequency (Figure 2.123)

The blocked-rotor test-data is converted to values that would have been obtained if this test were performed at rated voltage. If V_b is the voltage at which blocked-rotor test is performed, I_b the current input, and P the power input, the values converted to rated voltage V would be

$$I_{sc} = I_b \frac{V}{V_b}$$

$$P_{sc} = P_b \left(\frac{V}{V_b}\right)^2$$

$$\cos \phi_{sc} = \frac{P_b}{\sqrt{3\, VI_{sc}}}$$

OC equal to I_{SC} at an angle ϕ_{SC} is now drawn. A and C are then two points on the current locus, of which AC is a chord. The line AL is drawn at right angles to OV through A. The point X, where the right-bisector of chord AC meets AL, is the centre of locus-circle and XA is the radius.

CF, which is equal to $I_{sc} \cos \phi_{sc}$, represents to a certain scale, the power input to the motor $\sqrt{3}V\, I_{SC}$ (If on the current scale 1 cm = x amperes, and V is the rated line voltage, the power scale is 1 cm $\sqrt{3}\ V_x$ watts, for a 3-phase motor). The power intake under blocked-rotor conditions is consumed as no load loss (assumed constant) and stator copper loss and rotor copper loss. EF, which is equal to $I_0 \cos \phi_0$, represents on the power scale. No-load loss $\sqrt{3}\ VI_0 \cos \phi_0$. The balance of the power equal to CE is, which is the copper loss. If D divides CE such that CD/DE = where is the rotor resistance referred to stator copper loss at standstill. Line AC is termed as the output line, and line AD the torque line. For any operating point. P on the circle diagram, the entire performance may be obtained by drawing a perpendicular PT. For this operating condition.

Input Current	= OP
Power Output	= PQ
Rotor Copper loss	= QR

Stator Copper loss = RS

Friction, Windage and Iron Losses = ST

Power Input = PT

Torque in Synchronous Watts = PR

Slip = rotor copper loss/rotor input = $\dfrac{QR}{PR}$

Power Factor = $\dfrac{PT}{0P}$

Experimental Details

(a) For the no-load and blocked-rotor test, the motor will have to be connected through an induction regulator or three phase continuously variable auto-transformer to obtain variable voltages. Therefore, it is not necessary to connect the motor through a starter. Even when a starter is connected at voltage lower than the rated voltage, the armature of the no-volt coil will not hold, and the starter handle may have to be kept at the 'ON' position manually. It may be noted that in case starter is not connected in circuit, some other from of overload protection must be provided. (This may take the from of suitable fuses in the supply circuit).

It is desired to insert a starter in the circuit, it must be connected in the following manner.

(i) **Y-Δ starter :** After the measuring instruments (motors provided with Y-Δ starters are connected in Δ for normal operation).

(ii) **Auto-transformer starter :** Before the measuring instruments.

(b) Measurement of power.

Procedure

1. Connect as shown in Figure 2.124. Increase the voltage gradually till the motor starts. Record current, power input, applied voltage.

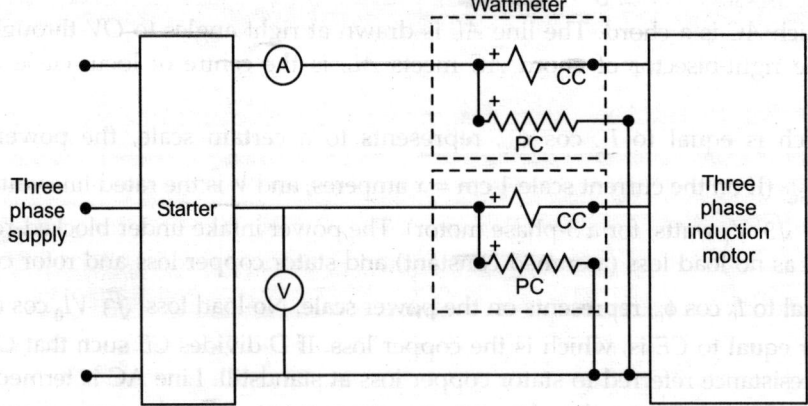

FIGURE 2.124. *Connection diagram of no-load and blocked rotor tests*

2. Increase the applied voltage in suitable steps, and record input current and power for various values of applied voltage up to about 125% of rated voltage.

3. Block the rotor, apply a reduced voltage (say 20% of the rated voltage) and record the voltage applied, current and power input.

4. Gradually increase the applied voltage, with rotor kept blocked and record 3 or 4 sets as in steps 3, till the stator current is about 1.2 times the rated current. These observations should be taken quickly to avoid

   (i) Overheating of the machine as the cooling is very poor when rotor is blocked,

   (ii) any errors in the observations on account of winding resistances changing with temperature.

5. Measure the resistance per phase (R_{ph}) of the stator.

If the motor is permanently connected in delta, and R_t is the resistance measured between any two terminals, then $R_{ph} = (3/2) R_t$. If the motor is connected in star and R_t is the resistance measured between any two line terminals, then

$$R_{ph} = \frac{1}{2} R_t$$

Results

1. Plot the variation of no-load power-input, current and power-factor as a function of voltage. Mark the points corresponding to rated voltage.

2. Plot the blocked-rotor power input, current, and power-factor as functions of applied voltage. From these curves find power input and necessary applied voltage corresponding to rated current at short circuit, and convert these values (power input, input current) for rated voltage.

3. From the data obtained in parts 1, 2 of the report draw the circle diagram and derive there from the values of current, power-factor, efficiency and slip corresponding to full-load. Also compute the (i) best power-factor (ii) maximum torque and starting torque as a percentage of full-load torque.

Discussion

Circle diagram, though a very convenient means of obtaining the induction motor performance, is based on certain assumptions which are not strictly true. Some of the approximations of circle diagram are discussed below.

1. The effective resistance of the rotor R_2 is not constant, but it changes with speed. At standstill, the rotor frequency is supply frequency. At this frequency, the skin effect (i.e., non-uniform distribution of current due to leakage flux) is predominant and the ratio $R_{effective}/R_{d.c.}$ (effective resistance/d.c resistance) is large, actual value of the ratio depending upon conductor configuration. As the rotor speed increases, the rotor frequency reduces and ratio $R_{effective}/R_{d.c.}$ also reduces. At normal running speeds skin effect is more or less absent and the effective resistance is very nearly equally to the d.c. resistance. In the circle diagram, it is assumed that rotor resistance remains constant. The circle diagram in which the rotor resistance is calculated from blocked-rotor test-data, therefore, gives too high rotor loss in the normal load range. The effort is appreciable only in the case of large motors employing deep conductors.

2. As speed changes rotor frequency also changes, which results a change in the rotor leakage reactance. But in circle diagram it is assumed to be constant.

3. The no-load losses are assumed constant. Over the normal operating speeds of the motor, frequency is low and, therefore, the rotor iron-losses are negligibly small. Under blocked-rotor conditions, friction and windage loss is zero but the rotor iron-losses become appreciable. It is assumed in the circle diagram that the sum of these losses (viz. stator iron-loss, rotor iron-loss, friction and windage losses and copper losses corresponding to the no-load component of current) is constant, though the individual component may vary.

4. It is assumed in the circle diagram that the power transferred across the gap is the rotor copper-loss and the mechanical power output. In fact, the rotor iron-losses and the friction and windage losses are also supplied out of the power transferred across the gap.

5. Another basic assumption of the circle diagram is that the mutual flux remains constant under all conditions, whereas actually it is not so.

2.9.4 Review Questions

1. Form the no-load test data, how can the bearing friction and windage losses be computed?
2. An induction motor intended for delta connection in normal operation? Why?
3. The rotor core-loss of an induction motor, when running, is usually neglected. Why?
4. Why does induction motor draw a higher magnetizing current than a transformer? Explain in brief.

PROGRAM. *Write Matlab program for plotting the torque versus speed characteristics of induction motor for variable frequency control keeping v/f ratio constant.*

2.9.5 Matlab Program to Draw Speed–torque Characteristics of Induction Motor

```
%****************************************************************
% Tw_im_vf.m - Plots developed torque(Nm) vs speed(rpm) for variable frequency
%****************************************************************
clear all; clf;
% Equivalent circuit parameters
R1=0.2; R2=0.3; X1=0.754; X2=0.754;
Rc=110; Xm=33.9;
p=4; fR=50; VR=460;
nmLim=2100; % nmLim is max allowable speed
% Thevenin resistance & inductance
Zm=j*Xm*Rc/(Rc+j*Xm);
ZTh=Zm*(R1+j*X1)/(R1+j*X1+Zm);
RTh=real(ZTh); LTh=imag(ZTh)/2/pi/fR;
L1=X1/2/pi/fR; L2=X2/2/pi/fR; Lm=Xm/2/pi/fR;
% Frequencies for analysis in addition to rated
nf=5; f=linspace(nmLim/nf,nmLim,nf)*p/120;
```

```
% Rated frequency calculations( Base curve )
npts=50; ws=2/p*2*pi*fR; wm=linspace(0,ws-0.01,npts);
w=2*pi*fR; vTh=abs(VR/sqrt(3)*Zm/(R1+j*X1+Zm));
for i=1:npts
   s=(ws-wm(i))/ws;
   TTd(i)=3*VTh^2*R2/s/ws/(  (RTh+R2/s)^2+w^2*(LTh+L2)^2  );
end
plot(wm*30/pi,TTd);grid;
title('Induction motor torque-speed for variable frequency');
xlabel('Speed, rpm'); ylabel('Total developed torque, N-m');
fdev=f(2)-f(1);
text(0.65,0.95,['Frequency increment: ',num2str(fdev),' Hz'],'sc')
text(0.65,0.92,['Rated Freq Curve(solidline): ',num2str(fR),' Hz'],'sc')
hold on; % Hold base curve for overplotting
for k=1:nf % Other than rated frequency calculations
ws=2/p*2*pi*f(k); wm=linspace(0,ws-0.01,npts);
w=2*pi*f(k);
% Set voltage-frequency control
if f(k)/fR <= 1; b=0.06*VR; VL=(VR-b)*f(k)/fR+b; else; VL=VR; end
% Empirical adjustment of core loss resistance
Rcf=0.5*Rc*(VR/VL*f(k)/fR)^2*(fR/f(k)+(fR/f(k))^2);
Zm=j*w*Lm*Rcf/(Rcf+j*w*Lm);
ZTh=Zm*(R1+j*w*L1)/(R1+j*w*L1+Zm);
RTh=real(ZTh); LTh=imag(ZTh)/2/pi/f(k);
VTh=abs(VL/sqrt(3)*Zm/(R1+j*w*L1+Zm));
for i=1:npts
   s=(ws-wm(i))/ws;
   TTd(i)=3*VTh^2*R2/s/ws/(  (RTh+R2/s)^2+w^2*(LTh+L2)^2  );
end
% Determination of points above nmmax for plot
smax=R2/sqrt(R1^2+w^2*(LTh+L2)^2); wmmax=(1-smax)*ws;
for m=1:npts; if wm(m)>=wmmax; break; end; end
plot(wm(m:npts)*30/pi, TTd(m:npts),'-');
end
disp('Press any key to see load curve');
pause;
TL=50+0.004052847*wm .^2; % superimpose load torque Charactersitc
plot(wm*30/pi, TL,'-.');
hold off;
%*************************************************************
```

PROGRAM OUTPUT. *The results of above written Matlab program is shown in Figure 2.125.*

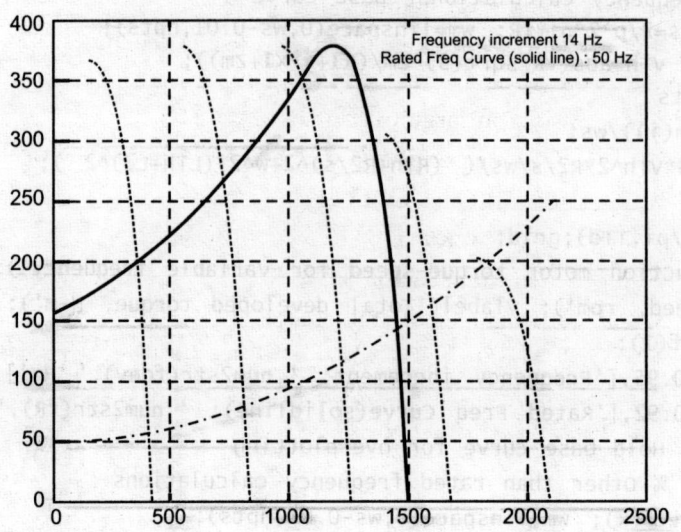

FIGURE 2.125. *Speed control of induction motor using variable frequency.*

2.9.6   Procedure of Induction Motor Design

Typical Specifications :

Output rating	–	kW
Voltage	–	Volts
Frequency	–	Hz
Speed	–	Rpm
Type of rotor	–	Cage/Slip ring
FL efficiency	–	%
Power factor	–	

Step 1. Main Dimensions

$$\text{Number of poles } P = \frac{120\,f}{N_s}$$

Where N_s = Synchronous Speed (rpm)

Average flux density in air gap B_{av} = 0.3 – 0.6T when a large over load capacity is required the B_{av} may be taken 0.65 T.

$$B_{av} = \text{Specific magnetic loading}$$

Choice of average flux density in air gap depends on the following :

1. The value of flux density in air gap should be small as otherwise the machine will draw a large magnetizing current giving a poor power factor.

2. An increased value of gap density results in increased iron loss.

3. It also depends upon the overload capacity of the machine.

Choice of ampere conductors (ac) per meter depends on the following :

1. The large value of ac means a greater amount of copper used in machine, which results higher copper losses and large temperature rise.
2. Small value of ac should be taken for high voltage machines, so space available for insulation should be more.
3. Large value of ac results in large overload capacity.

Ampere conductor per meter of periphery $ac = 5000\text{-}45000 = $ Sp. electrical loading

Output Equation $Q = Co\ D^2 L\ Ns$

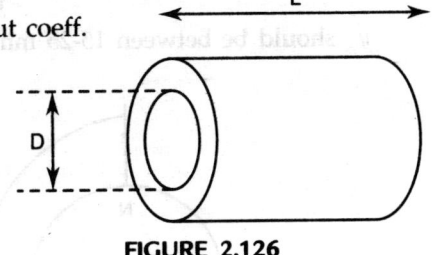

FIGURE 2.126

where
$Co = 11\ Bav\ Kw\ ac \times 10^{-3} = $ output coeff.
$Kw = $ winding factor (0.955)
$D = $ Diameter of stator bore
$L = $ Length of machine
$N_s = $ synchronous speed in rps

$$D^2 L = \frac{Q}{CoN_s} \qquad \text{...(1)}$$

Select the ratio L/τ

$$\tau = \text{Pole Pitch} = \pi D/p$$

For	Value of L/τ
Minimum Cost	1.5–2
Good pf	1.0–1.25
Good h	1.5
Good overall design	1.0

$$\frac{L}{\tau} = \frac{LP}{\pi D} \qquad \text{...(2)}$$

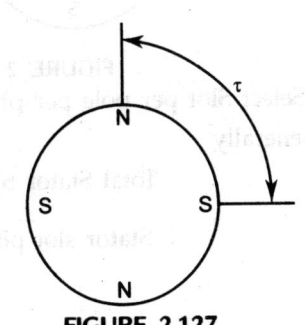

FIGURE 2.127

From equations (1) & (2) calculate L and D.

Calculate peripheral speed $N_p = \dfrac{\pi DN}{60}$

N_p should be less than 30 m/s, but for special rotor construction it may be taken upto 75 m/s. If peripheral speed exceeds the above limits then reduce it to 30 m/s for small or medium kW machines and 75 m/s for high power machine. Do back calculations and calculate 'D'.

$$D = \frac{60\ N_p}{\pi N} \quad \text{where } N = \text{rpm}$$

using the modified value of 'D' calculate 'L' using equation (1).

Modified pole pitch $\tau = \dfrac{\pi D}{P}$

Net iron Length $L_i = 0.9\ L$

0.9 is the stacking factor. Generally 0.5 mm thick Lamination is used.

Step 2. Stator Design

Calculate Stator Voltage per Phase = E_s depending upon Y/Δ connection of stator winding.

Flux per pole $\phi m = B_{av}\, \tau\, L$

Stator turns per phase $T_s = \dfrac{E_s}{4.44\, f\, \phi_m\, K_w}$

K_w = Winding factor (0.955)

y_{ss} = Stator Slot Pitch

y_{ss} should be between 15-25 mm but can be less than 15 mm for semi-enclosed slots.

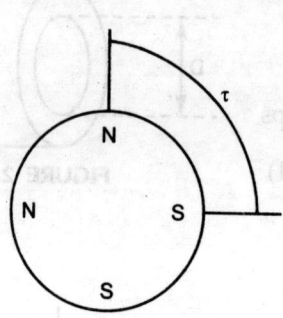

FIGURE 2.128

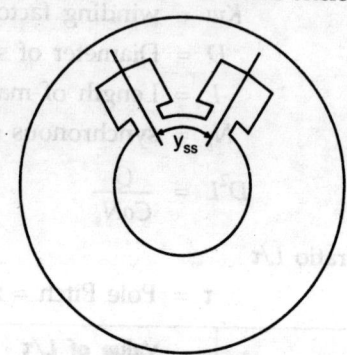

FIGURE 2.129

Select Slot per pole per phase q_s = 2,3,4 ....

Generally q_s = 2 or 3

Total Stator Slots $S_s = 3pq_s$

Stator slot pitch $y_{ss} = \dfrac{\pi D}{S_s}$

P = No. of poles

Total Stator Conductors = $6\, T_s$

Stator conductor per slot $Z_{SS} = \dfrac{6T_s}{S_s}$

Step 3. Conductor Size

Stator current per phase $I_s = \dfrac{kW \times 10^3}{3E_s \times \eta \times pf}$

Stator line current = $\sqrt{3} I_s$, For Δ connected stator winding

Stator line current = I_s, For Y connected stator winding

Choose current density in stator conductor $\delta s = 3.5 - 3.4$ A/mm^2

From table 2.6 the nearest standard conductor size should be selected.

Diameter of the bare conductor $d = 2\sqrt{(a_s / \pi)}$

$a_s = I_s/\delta_s$ mm^2

Modify the area of conductor $a_s = \pi d^2/4$

Current density in stator conductor $\delta_s = I_s / a_s$

Select the diameter of enameled conductor, d_i from Table 2.6 for medium covering.

TABLE 2.6. Round Copper Wire (Synthetic Enamel)

Nominal Conductor Diameter (mm)	Overall diameter			
	Fine covering (mm)	Medium covering (mm)	Thick covering (mm)	Extra thick covering (mm)
0.050	0.060	0.065	-	-
0.060	0.073	0.078	-	-
0.070	0.084	0.092	-	-
0.080	0.095	0.105	0.118	-
0.090	0.105	0.115	0.128	-
0.100	0.118	0.128	0.141	-
0.112	0.132	0.143	0.156	-
0.125	0.146	0.159	0.172	-
0.132	0.155	0.168	0.181	-
0.140	0.163	0.176	0.191	-
0.150	0.173	0.186	0.201	-
0.160	0.185	0.198	0.213	-
0.170	0.198	0.211	0.226	-
0.180	0.208	0.223	0.238	-
0.195	0.224	0.239	0.257	-
0.200	0.230	0.246	0.264	-
0.212	0.242	0.258	0.276	-
0.224	0.254	0.272	0.290	-
0.236	0.266	0.284	0.302	-
0.250	0.283	0.301	0.319	0.344
0.258	0.291	0.309	0.327	0.352
0.265	0.299	0.317	0.335	0.362
0.280	0.316	0.334	0.351	0.379
0.300	0.336	0.354	0.372	0.400
0.307	0.345	0.362	0.380	0.408
0.315	0.354	0.372	0.389	0.417
0.335	0.375	0.393	0.410	0.441
0.355	0.397	0.415	0.432	0.465
0.375	0.418	0.438	0.455	0.489
0.400	0.445	0.465	0.483	0.518
0.425	0.472	0.493	0.513	0.549
0.462	0.511	0.531	0.551	0.588
0.475	0.526	0.546	0.566	0.603
0.500	0.551	0.571	0.591	0.630
0.530	0.581	0.602	0.623	0.662
0.560	0.612	0.635	0.655	0.696
0.600	0.654	0.677	0.697	0.738
0.630	0.684	0.707	0.728	0.768
0.670	0.727	0.750	0.771	0.812
0.710	0.768	0.791	0.814	0.852
0.730	0.788	0.811	0.834	0.874
0.750	0.808	0.831	0.854	0.895
0.800	0.861	0.884	0.907	0.950
0.850	0.912	0.935	0.958	1.001

0.925	0.990	1.016	1.039	1.085
0.950	1.015	1.041	1.064	1.110
1.000	1.070	1.095	1.120	1.165
1.060	1.130	1.155	1.180	1.225
1.120	1.190	1.215	1.240	1.287
1.180	1.253	1.278	1.303	1.353
1.250	1.325	1.350	1.375	1.425
1.320	1.395	1.420	1.447	1.500
1.400	1.480	1.505	1.535	1.585
1.500	1.580	1.605	1.635	1.685
1.600	1.680	1.710	1.740	1.790
1.700	1.785	1.810	1.840	1.890
1.800	2.885	1.915	1.940	1.995
1.900	2.990	2. 015	2.045	2.095
2.060	2.150	2.180	2.210	2.265
2.120	2.211	2.241	2.271	2.327
2.240	2.335	2.365	2.347	2.455
2.360	2.486	2.488	2.420	2.575
2.500	2.600	2.630	2.665	2.720
2.650	2.752	2.785	2.815	2.872
2.800	2.905	2.935	2.970	3.025
2.900	3.010	3.040	3.070	3.125
3.00	3.110	3.140	3.175	3.230
3.150	3.262	3.295	3.325	3.382
3.250	3.365	3.395	3.427	3.485
3.350	3.465	3.497	3.530	3.585
3.450	3.567	3.600	3.630	3.690
3.550	3.670	3.700	3.732	3.795
3.650	3.770	3.800	3.835	3.895
3.750	3.872	3.902	3.937	3.997
4.00	4.125	4.155	4.190	4.255

Step 4. Slot Dimensions

Space required for bare conductors in a slot = $Z_{ss} \times a_{ss}$

Space factor for slot = area of copper conductor slot/
Area of slot

$$= 0.25 - 0.4$$

Minimum allowable flux density in stator tooth = 1.7 T

Minimum width of stator teeth (Wts) min = $\dfrac{\phi_m P}{1.7 \times Ss \times Li}$

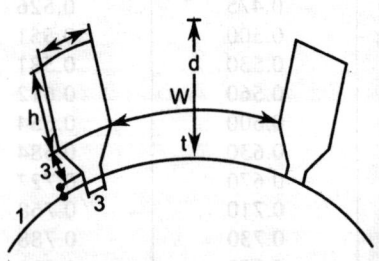

FIGURE 2.130

Area of stator slot $A_s = Z_{ss}\, a_{ss} \times$ space factor

$S_s\,[AB + (Wts)\,min)] = \pi(D + 8)$, where D in mm

Slot width $W_{ss} = AB = \dfrac{\pi(D+8)}{S_s} - (Wts)\,min$

Calculate d_{ss} and $h(= dss{-}4)$ can be calculated.

Length of mean turn, $L_{mts} = 2L + 2.3t + 0.24$ meters

Step 5. Stator Teeth

Tooth width varies according to height. Average flux density in stator tooth is calculated at $1/3^{rd}$ height form narrow end.

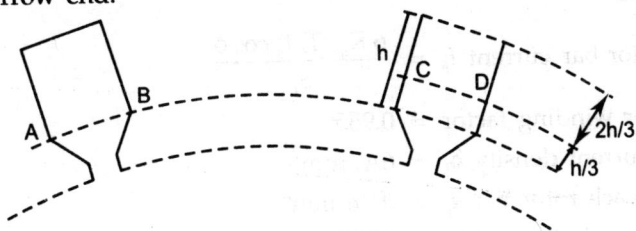

FIGURE 2.131

Width $CD = \dfrac{\pi(D + 8 + 2h/3)}{S_s} - W_{ss} = \dfrac{W_h}{3}$

Flux density in stator teeth $= \phi_m p/(Ss\; W_{h/3}\; Li)$ wb/m2

Step 6. Stator Core

Flux in stator core $= \phi_m/2$

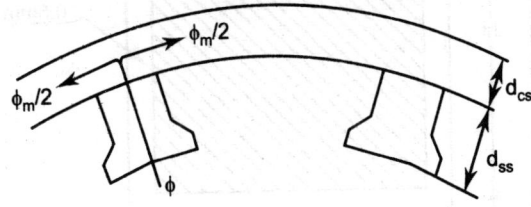

FIGURE 2.132

Assume flux density 1.2 Wb/m^2 in stator core.

Area of stator core $Acs = \dfrac{\phi_m}{(2x\,1.2)}$

$dcs = \dfrac{Acs}{Li}$

Make a round figure of core depth dcs and again calculate flux density in stator core

Outer diameter of stator $Do = D + 2d_{ss} + 2d_{cs}$

$Bcs = \dfrac{\phi_m}{(2 \times dcs \times Li)}$

Step 7. Rotor Design

Air gap length $l_g = (0.2 + 2\sqrt{(LD)})$ mm, where D – meter and L – meter.

Diameter of rotor $D_r = D - 2\,l_g$

Step 8. Rotor Slots

Number of rotor slots (S_r) should be selected that

$$Ss - Sr \neq 0, \pm p, \pm 2p, \pm 3p, \pm 5p, \pm 1, \pm 2, \pm(p \pm 1), \pm (p \pm 2)$$

For $S_s - S_r = 0, 3p$ magnetic locking occurs

For $S_s - S_r = p, 2p, 5p$ synchronous cusps (crawling).

For $S_s - S_r = 1, 2, (p \pm 1), (j \pm 1)$ Noise and vibration appears.

Rotor slot pitch at air gap = $ysr = \dfrac{\pi D_r}{S_r}$

Step 9. Rotor Bars

$$\text{Rotor bar current } I_b = \frac{6 K_{ws} T_s I_s \cos \phi}{S_r}$$

where K_{ws} – Stator winding factor = 0.955

Take rotor bar current density δ_b = 6A/mm^2

Area of each rotor bar a_b = Ib/δ mm^2

Slot dimensions are shown in Figure 2.133.

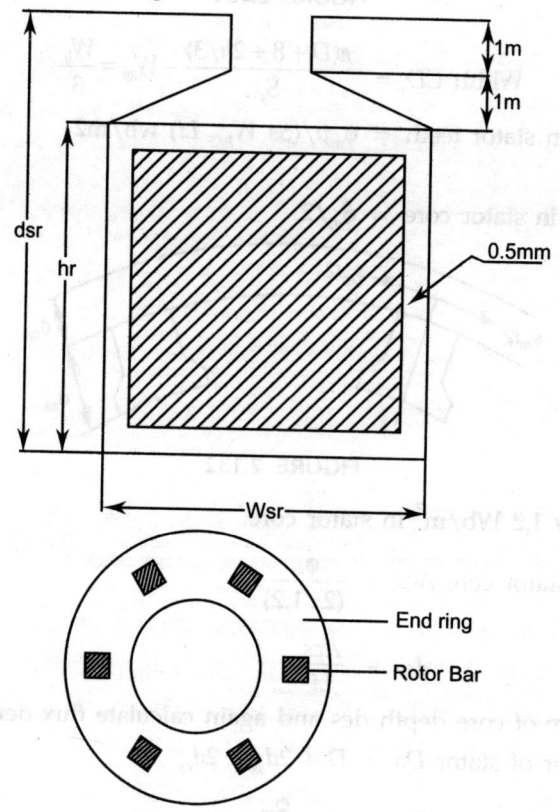

FIGURE 2.133

$$hr = (dsr - 4) \text{ mm}$$

Slot pitch at bottom of rotor slot = $\dfrac{\pi(Dr - 2dsr)}{Sr}$

Tooth width at rotor W_t = Slot pitch at bottom – W_{sr}

Flux density at bottom of rotor tooth

$$= \frac{\varphi_m P}{W_{sr\,1/3} \times Sr \times Li} < 1.7 \text{ Tesla}$$

If flux density in rotor teeth is greater than 1.7 T then adjust it by choosing another cross section of rotor bar.

$$\text{Length of rotor bar} = 1.1 \times L \text{ (due to end rings)}$$

$$= L_b$$

$$Wsr_{1/3} = \pi[Dr - 2dsr + 2\, hr/3] - Wsr$$

$$\text{Resistance of each bar } r_b = \frac{0.021 \times L_b}{a_b}$$

Total cu loss in bars $= Sr \times I_b^2 r_b = W_{cu\text{-}rotor}$

Step 10. End Rings

$$\text{End ring current } I_e = \frac{Sr \times I_b}{p\pi}$$

Current density in end ring $\delta_e = 6\text{A/mm}^2$

Area of end ring $a_e = I_e/\delta_e$

Select from table 2.6 the cross section area and dimensions of conductor for end rings

Outer diameter of end ring, $D_{outer} = (Dr-4)$ mm

Inner diameter of end ring, $D_{inner} = (Dr-2dsr)$ mm

Mean diameter of end ring $De = (D_{outer} + D_{inner})/2$

Copper loss in two end rings $= W_{cu\_end\_ring}$

$$W_{cu\_end\_ring} = 2 \times r_e \times I_e^2$$

$$\text{Resistance of each end ring } r_e = \frac{0.021 \times \pi D_e}{a_s}$$

Total cu loss in rotor $= W_{cu\_rotor} + W = W_{cu\_end\_ring}$

Rotor copper loss/rotor output $= S/(1 - S)$

Calculate slip 'S' should be less than 8%.

Step 11. Rotor Core

Flux density in rotor core $B_{cr} = 1.2 \text{ wb/m}^2$

$$\text{Depth in rotor core } d_{cr} = \frac{\phi_m}{2\, B_{cr}\, L_i}$$

Inner diameter of rotor lamination $D_i = D_r - 2d_{sr} - 2d_{cr}$

Step 12. No Load Current

Magnetizing Current

(i) *Air gap:* Calculate (Slot opening (W_o)/air gap length (l_g)) in stator and rotor both and find corresponding Carter's coefficient for these ratios from Figure 2.134.

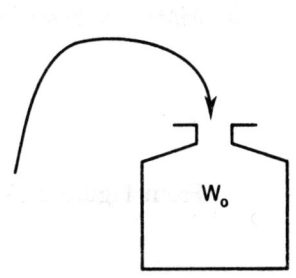

FIGURE 2.134

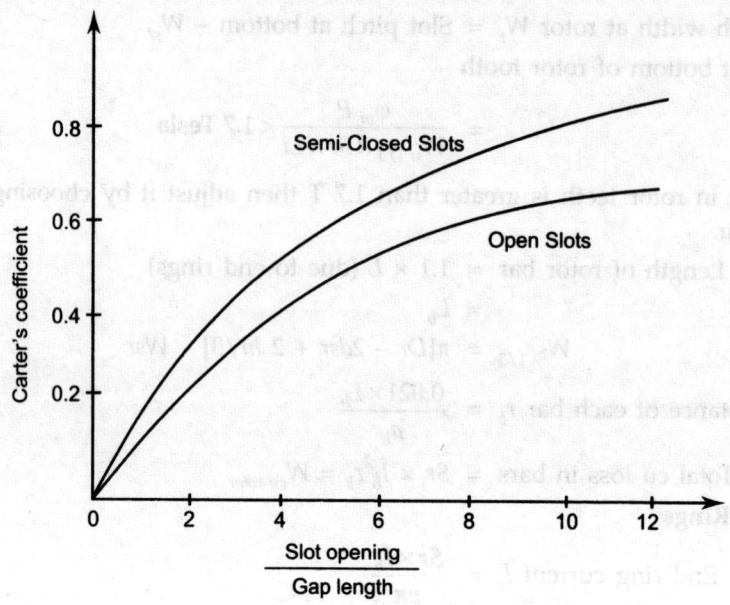

FIGURE 2.135. *Carter's air gap coefficient*

Gap contraction factor for stator slot $K_{gss} = \dfrac{yss}{(yss - KcsWo)}$

Net gap contraction factor for slots

$$K_{gs} = K_{gss} \times K_{gsr}$$

$$B_{g60} = 1.36\, B_{av}$$

Effective length of air gap $l_{ge} = K_{gs} \times l_g$

MMF for air gap $AT_g = 800000\, B_{g60}\, l_{ge} \to$ meter

(ii) *MMF For Stator Teeth:* Flux density at one third height from narrow end

$$B_{ts1/3} = \frac{\phi_m P}{W_{h/3} \times Ss \times Li}$$

$$B_{ts1/360} = 1.36\, B_{ts1/3}$$

Find its ATts from AT–Flux density curve

MMF required for Stator teeth $AT_{ts} = atts \times d_{ss}$

(iii) *MMF For Rotor Teeth:* Flux density in rotor teeth at one third height from narrow end

$$B_{tr1/3} = \frac{\phi_m P}{W_{sr1/3} \times Sr \times Li}$$

$$B_{tr60} = 1.36 \times (B_{tr1/3})$$

From Figure 2.136 find attr MMF required for rotor teeth $AT_{tr} = attr \times d_{sr}$

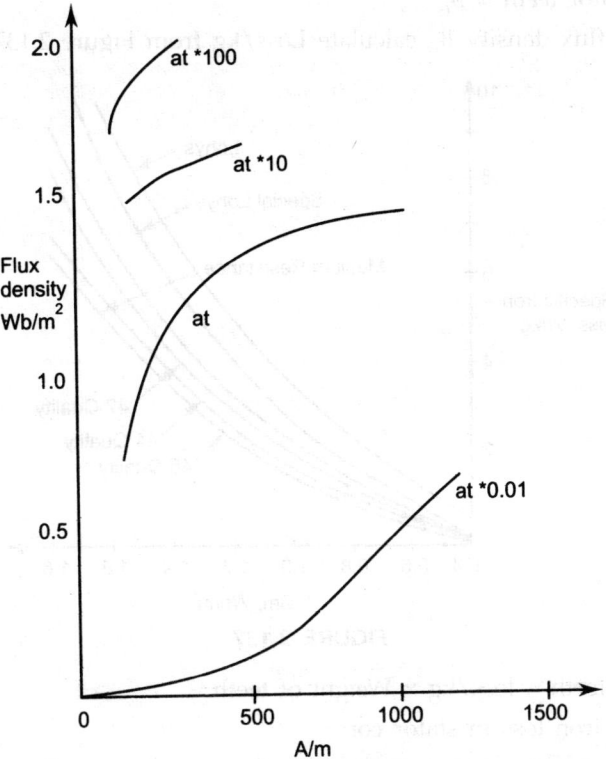

FIGURE 2.136. B-H curve for electrical steel (non-oriented)

(iv) *MMF For Stator Core:* Length of magnetic path through air gap in core stator

$$Lcs = \frac{\pi(D + 2dss + dcsr)}{3p}$$

Flux density in stator core = B_{cs}

Find atcs from curve

MMF for stator core per pole $AT_{cs} = l_{cs} \times at_{cs}$

(v) *MMF For Rotor Core:* Flux density in rotor core = B_{cr}

Find atcr from curve $Lcr = \frac{\pi(D_r + 2d_{sr} + d_{cr})}{3p}$

MMF for rotor core $AT_{cr} = l_{cr} \times at_{cr}$

Total magnetizing mmf per pole $AT_{60} = At_g + AT_{ts} + AT_{tr} + AT_{cs} + AT_{cr}$

Magnetizing current per phase

$$I_m = \frac{0.427 \, pA \, T_{60}}{K_{ws} T_s}$$

Where $K_{ws} = 0.955$

Step 13. Loss Component

Calculate volume of stator teeth in m^3

Weight of stator teeth = $7.6 \times 10^3 \times$ Volume

Flux density in stator teeth = B_{ts}

Corresponding to flux density B_{ts} calculate Loss/kg from Figure 2.137.

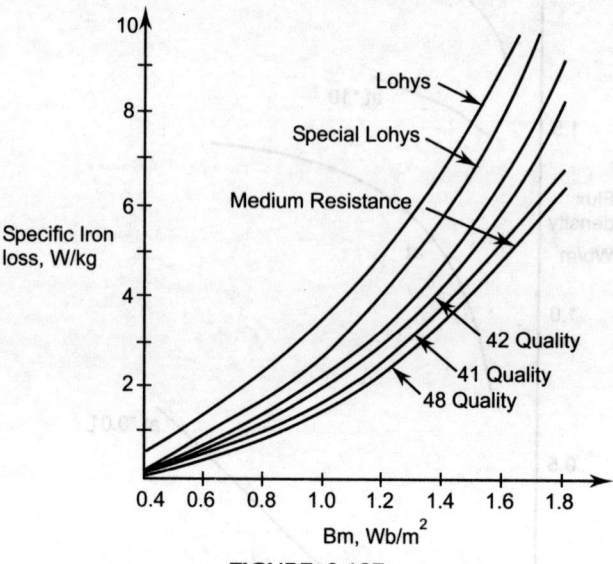

FIGURE 2.137

Iron loss in stator teeth = loss/kg × Weight of teeth

Similarly calculate iron loss in stator core.

Iron loss on stator side = iron loss in stator teeth + iron loss in stator core

Total iron loss = 2 × iron loss on stator side.

Friction and Windage losses = 3% of output

Total no load loss = Friction and Windage losses + Total Iron losses

Loss component of no load current $I_i = \dfrac{\text{Total no load losses}}{(3 \times \text{Voltage per phase})}$

Step 14. Short Circuit Current

No – Load current $I_o = \sqrt{(I_i^2 + I_m^2)}$

$$\lambda ss = \mu_o \left[\frac{h_1}{3W ss} + \frac{h_2}{W ss} + \frac{2h_3}{(W ss + W o)} + \frac{h_4}{W o} \right]$$

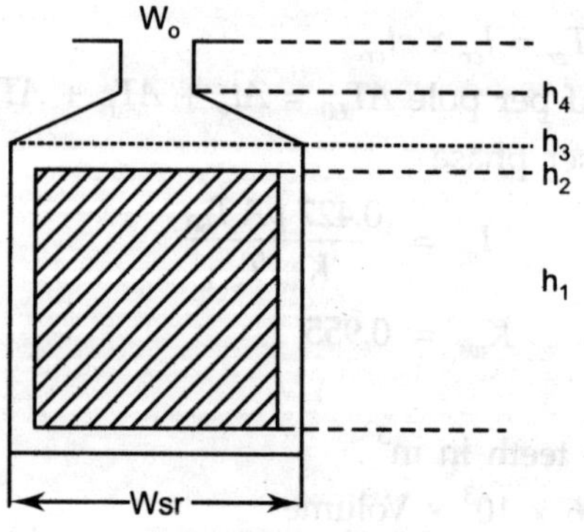

FIGURE 2.138

Power factor of no load current $\cos \phi_o = I_i/I_o$

Leakage Reactance for stator slot-

Sps - slot permeance for stator slot-

Spr - slot permeance for rotor slot-

$$\lambda sr = \mu_o \left[\frac{h_1}{3Wsr} + \frac{h_2}{Wsr} + \frac{2h_3}{(Wsr + Wo)} + \frac{h_4}{Wo} \right]$$

This referred to stator side $\lambda_{sr'} = \lambda_{sr} K_{ws}^2 S_s /(K_{wr}^2 S_r)$

Total Sp. Slot performance -

      Slot leakage reactance $X_s = 8 \pi f T_s^2 L \lambda_s/(pq)$

Overhang Leakage

                  Coil Span = Slots/Pole should be an odd integer if even then

                  Coil Span = Slot/Pole – 1

                  $\lambda_s = \lambda_{ss} + \lambda_{sr'}$

Coil Span / Pole Pitch (Slots) = Coil Span / Slot per Pole

Refer Figure 2.139 corresponding to above ratio find its Ks from curve.

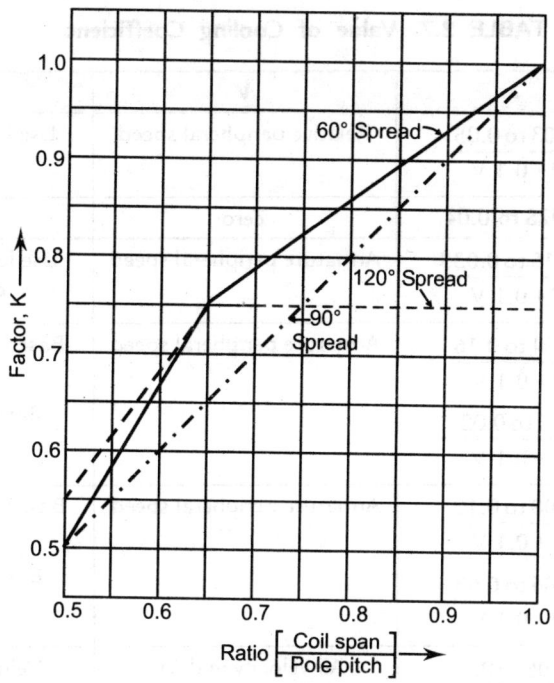

FIGURE 2.139

Specific Overhang Permeance $L_o \lambda_o = \mu_o K_s t^2/(\mu \, y_{ss})$

Zigzag Leakage

Magnetizing reactance $x_m = E_s /P_m$

Overhang leakage reactance $x_o = 8 \pi f T_s^2 L_o \lambda_o/(pqs)$

Zigzag leakage reactance per phase $X_z = (5/6) (x_m/9) (1/q_s^2 + 1/q_r^2)$

Total leakage reactance per phase = Slot leakage reactance + Overhang leakage reactance
+ Zigzag leakage reactance

Resistance

Resistance of stator winding per phase $r_s = 0.021\ T_s\ L_{mts}\ /a_s$

Total stator cu loss = $3\ I_s^2\ r_s$

Rotor resistance refer to stator r_r = Rotor copper loss/Is^2

Total resistance referred to stator per phase $R_s = r_s + r_r'$

Impedance $Z_s = \sqrt{(R_s^2 + X_s^2)}$

Short circuit current $I_{sc} = E_s\ /Z_s$

Short circuit power factor $Cos\ \phi_{sc} = R_s/Z_s$

η = Output /(output +Losses)

STATOR TEMPERATURE RISE

Outer diameter of stator $D_o = D + 2(d_{ss} + d_{cs})$

Outer cylindrical surface of stator $S_f = \pi\ D_o\ L$

Cooling coefficient $C_1 = 0.03$ calculated from the Table 2.7:

TABLE 2.7. Value of Cooling Coefficient

Part	c	V	Remarks
Cylindrical surface of stator and rotor	$\dfrac{0.03\ to\ 0.05}{1+0.1\ V}$	Relative peripheral speed	Use low values for forced cooling
Back of stator core	0.025 to 0.04	Zero	
Cylindrical surface of d.c. armatures	$\dfrac{0.015\ to\ 0.035}{1+0.1\ V}$	Armature peripheral speed	Use lower values for large open machines
Stationary field coils	$\dfrac{0.14\ to\ 0.16}{1+0.1\ V}$	Armature peripheral speed	Based on total coil surface
	$\dfrac{0.0\ to\ 0.08}{1+0.1\ V}$		Based on exposed coil surface
Rotating field coil	$\dfrac{0.08\ to\ 0.12}{1+0.1\ V}$	Armature peripheral speed	Based on total coil surface
	$\dfrac{0.06\ to\ 0.03}{1+0.1\ V}$		Based on exposed coil surface
Ventilating ducts	$\dfrac{0.08\ to\ 0.2}{V}$	Air velocity in ducts	Velocity in ducts is 10 percent of peripheral speed of core
Commutator	$\dfrac{0.015\ to\ 0.025}{1+0.1\ V}$	Commutator peripheral speed	

Loss dissipation from back of core

$$L_1 = \pi D_o\ L\ /0.03\ \text{w/}^\circ\text{c}$$

Inside cylindrical surface of stator = $\pi D L$

Peripheral speed $V_a = \pi D n_s$ where, n_s = rps

Cooling coefficient = $0.04 / (1 + 0.1 \, Va) = C_2$

Loss dissipation from inner surface of stator

Velocity of air at end surface = $0.1 \, V_a$

coefficient $C_3 = 0.15/(Va \times 0.1)$

$L_2 = \pi D_l/$Cooling coeficient C_2

Cooling surface of two ends = $2 \times \pi(Do^2 - D^2)/4$

Loss dissipation from end surface

L_3 = (Cooling surface of two ends/Cooling coefficient C_3) w/ $^\circ$c

Total Loss dissipated = $(L_1 + L_2 + L_3)$

Temperature rise θm = Total stator losses/$(L_1 + L_2 + L_3)$

2.9.7   Matlab Program for Induction Motor Design

```
%% input data
kw=input('output in kw=');
m=input('phases=');
vs=input('supply voltage=');
default=input('Remaining information as default - press 1\n otherwise
    press - 0 ');
if (default ==1)
    f=50; %input      ('supply frequecny=');
    p=4; %input       ('number of poles=');
    nr=1400; %input ('rotor speed=');
    connection=2;     %input('conection of stator winding(1 for star and
                        2 for delta)=');
    effi=0.82;        %input('efficiency=');
    pf=0.82;          %input('power factor=');
    q=2;              %input('slots per pole per phase(2 or 3)=');
    ssf=0.3;          %input('stator slot space factor(between 0.25 and 0.40)=');
    kws=0.9;          % stator winding factor
    Bav=0.40;
    ac=20000;
else
    f=input('supply frequecny=');
    p=input('number of poles=');
    nr=input('rotor speed=');
    connection=input('conection of stator winding(1 for star and 2 for delta)=');
    effi=input('efficiency=');
    pf=input('power factor=');
    q=input('slots per pole per phase(2 or 3)=');
    ssf=input('stator slot space factor(between 0.25 and 0.40)=');
    Bav=input('Specific Magnetic Loading');
    ac=input('Specific Electrical Loading');
end
```

```
if connection==1 ;
   vph=vs/1.732;
else
   vph=vs;
end;

kva=kw/(effi*pf);
ns=2*f/p; %in rps

c0=11*Bav*ac*kws*(10.^(-3));
D2L=kva/(c0*ns);
LbyD=(3.1416/p); %for good overall design
Dcube=D2L/LbyD;
D=Dcube.^(1/3);
L=LbyD*D;
peri_speed=3.1416*D*ns; % in meter/sec.

if peri_speed >=55;
   D=55*ns/3.1416;
end;
toy=3.1416*D/p;
fluxpp = Bav*toy*L; %flux per pole
Ts=vph/(4.44*f*fluxpp*kws); %stator turns per phase
Ts=round(Ts);
Ts=2*(round(Ts/2)); % to make even integer

Iph=kva*1000/(3*vph);
delta_s=4 * (10^(6)); %stator current desity in amp per met sq
as= Iph/delta_s; % cross sectional area of stator conductor in met sq.
dia_st_cond=sqrt(4*as/3.1416);

Ss= m*p*q;
total_s_cond=m*2*Ts;
Zss=total_s_cond/Ss; %conductors per stator slot
Zss=round(Zss);
Zss=2*(round(Zss/2)); % to make even integer
Ts=Zss*Ss/(2*m); %modified

s_slot_area=Zss*as/ssf; %stator slot area
Wtsmin= (fluxpp*p)/(1.7*Ss*L);
Wtsmin=Wtsmin*1.20; %to aviod saturation in stator tooth
Wss=(3.1416*(D+2*3*0.001)/Ss)- Wtsmin;
dss=((s_slot_area*1000000-4)/(Wss*1000) +2)/1000; %meters
hss=dss-0.003;
Wts13=3.1416*(D+(0.003*2)+(2*hss/3))/Ss -Wss;
Bts13=p*fluxpp/(Ss*Wts13*L);
```

```
Acs=fluxpp/(2*1.2);
dcs= Acs/L;
dcs= (round(dcs*1000))/1000; %meters
Acs=dcs*L;
Bcs=fluxpp/(2*Acs);

%%%% ROTOR DESIGN
lg=0.2+2*(sqrt(L*D)); %air gap length in mm
lg=(round(lg*10))/10; %air gap length in mm
lg=lg/1000; %air gap length in meters
Dr=D-2*lg;
% selection of number of rotor slots
p=4;
diff=[0, 3*p, p, 2*p, 5*p, 1, 2, p+1, p-1, p+2, p-2];
smax=max(diff);
smin=min(diff);

[row,col]=size(diff);
rnum=smin;
for num=smin:smax
   for j=1:col
      if num==diff(j)
         rnum=num;
      end;
end;
   diffnum=num-rnum;
   if diffnum ~= 0;
   reqnum=num;

break;
end;
end;
Sr=Ss-reqnum;

%calculation of rotor bar dimensions
Ib=0.85*6*Ts*Iph/Sr;
ab=Ib/6000000; %area rotor bar in met sq.

Wtrmin=p*fluxpp/(Sr*L*1.7);
Wtrmin=1.1*Wtrmin;
aq=2*3.1416;
bq=Wtrmin*1000-4*3.1416-3.1416*Dr;
cq=ab*1000000+Wtrmin*1000-3.1416*Dr*1000-6*3.1416;
bh=(-bq+sqrt(bq*bq-4*aq*cq))/(2*aq); %rotor conductor bar height mm
bw=ab*1000000/bh;  %rotor conductor bar width mm
if bh <= 2.0*bw
   bw=(sqrt(ab*1000000/1.5)); % mm
   bh=2.0*bw; % mm
end
```

```
dsr=bh+3; % mm
hr=dsr-2.5; %mm
wsr=bw+1; %mm
Wtrmin=3.1416*(Dr*1000-2*dsr)/Sr - Wsr; % mm
Wtr13=3.1416*(Dr*1000-2*dsr+2*hr/3)/Ss -Wsr; % mm
Btr13=p*fluxpp/(Sr*Wtr13*0.001*L); % web/met sq.
%End rings
Ie=Sr*Ib/(3.1416*p);
ae=Ie/6000000; % met sq.
endringbh=1.2*bh; %mm
endringbw=ae*1000000/endringbh; % mm
Deout=Dr*1000-2*2-hr+endringbh;
Deinn=Dr*1000-2*2-hr-endringbh;
Demean=(Deout+Deinn)/2; %mm
Rend=0.021*(3.1416*Demean *0.001)/(ae*1000000); % ohms
Wcuend=Ie*Ie*Rend*2; %watts
Rbar=0.021*L/(ab*1000000); % ohms
Wcubar=Ib*Ib*Rbar*Sr;
Wcurotor=Wcubar+Wcuend; % watts
Wcurotorperphase=Wcurotor/3;
R_rotor_wrts=Wcurotorperphase*1*1/(kws*kws*Iph*Iph);
slip=Wcurotor/(kw*1000+Wcurotor);

Bcr=1.2;
dcr=fluxpp/(2*Bcr*L); % meters
Dinn_rot_lamin=Dr-2*dcr-2*dsr*0.001; % meters


%%%%magnetising current

%air gap
yss=3.1416*D/Ss;%met
ysr=3.1416*Dr/Sr;%met
wobygaps=2/(lg*1000);
x=wobygaps;

Kcs= -6.7813e-005*x.^4+0.0021*x.^3-0.0260*x.^2+0.1982*x.^1+0.0328;
%Kcs=0.7;
wobygapr=1/(lg*1000);
x=wobygapr;
Kcr= -6.7813e-005*x.^4+0.0021*x.^3-0.0260*x.^2+0.1982*x.^1+0.0328;
%Kcr=0.5;


wos=0.001*2;
wor=0.001*1;
Kgss=yss/(yss-Kcs*wos);
Kgsr=ysr/(ysr-Kcr*wor);
Kgs=Kgss*Kgsr;
Bg60=1.36*Bav;
```

```
lge=Kgs*lg;
ATgap=800000*Bg60*lge;


%stator teeth
Bts13_60=1.36*Bts13;
Bat=Bts13_60;
% Calculations of AT
if Bat <= 1.7
    atts= -7.6703* Bat.^15+30.1849*Bat.^14 -30.622*Bat.^13 +15.3884*Bat.^9 -
      15.9632*Bat.^3+26.8724*Bat-1.4687;
else
    atts= (0.4571* Bat.^15-2.0683*Bat.^14 +2.4958*Bat.^13 -1.4753*Bat.^10)*1000;
end;

ATts=atts*dss;

%stator core
Bcs_60=1.36*Bcs;
Bat=Bcs_60;
if Bat <= 1.7
    atts= -7.6703* Bat.^15+30.1849*Bat.^14 -30.622*Bat.^13 +15.3884*Bat.^9 -
      15.9632*Bat.^3+26.8724*Bat-1.4687;
else
    atts= (0.4571* Bat.^15-2.0683*Bat.^14 +2.4958*Bat.^13 -1.4753*Bat.^10)*1000;
end;
atcs=atts;
Lcs=3.1416*(D+2*dss+dcs)/(3*p); %meters
ATcs=atcs*Lcs;

%rotor teeth
Btr13_60=1.36*Btr13;
Bat=Btr13_60;
if Bat <= 1.7
    atts= -7.6703* Bat.^15+30.1849*Bat.^14 -30.622*Bat.^13 +15.3884*Bat.^9 -
      15.9632*Bat.^3+26.8724*Bat-1.4687;
else
    atts= (0.4571* Bat.^15-2.0683*Bat.^14 +2.4958*Bat.^13 -1.4753*Bat.^10)*1000;
end;attr=atts;
ATtr=attr*dsr/1000;

%rotor core
Lcr=3.1416*(Dr-2*dsr*0.001-dcr)/(3*p); %meters
Btr13_60=1.36*Btr13;
if Bat <= 1.7
    atts= -7.6703* Bat.^15+30.1849*Bat.^14 -30.622*Bat.^13 +15.3884*Bat.^9 -
      15.9632*Bat.^3+26.8724*Bat-1.4687;
else
    atts= (0.4571* Bat.^15-2.0683*Bat.^14 +2.4958*Bat.^13 -1.4753*Bat.^10)*1000;
end;
```

```
atcr=atts;
ATcr=atcr*Lcr;

AT60=ATgap+ATts+ATcs+ATtr+ATcr;
Imph=0.427*p*AT60/(kws*Ts);

%loss component

%STATOR TOOTH
vol_ts=(3.1416*((((D+2*dss).^2)/4)-((D.^2)/4))-Ss*s_slot_area)*L;
wght_ts=7600*vol_ts;
Bts13peak=(3.1416/2)*Bts13;
Biron=Bts13peak;
wattperkg= 5.3751*(Biron.^2)+2.3571*Biron-1.200 ;
%wattperkg=( 12 );
Wiron_ts=wattperkg*wght_ts;

%STATOR CORE
vol_cs=(3.1416*(D+2*dss+dcs))*dcs*L;
wght_cs=7600*vol_cs;
Bcspeak=(3.1416/2)*Bcs;
Biron= Bcspeak;
wattperkg= 5.3751*(Biron.^2)+2.3571*Biron-1.200 ;
Wiron_cs=wattperkg*wght_cs;

Wiron=2*(Wiron_ts+Wiron_cs);
Wfw=0.03*kw*1000;% F & W losses
Wnl=Wiron+Wfw; % total no load losses
Icph=Wnl/(3*vph);
I0=sqrt(Imph.^2 + Icph.^2);
I0percent=I0*100/Iph;

Lmts=2*L+2.5*toy;
R_stator=0.021*Lmts*Ts/(as*1000000);
Wcu_stator=3*Iph*Iph*R_stator;
losses_total=Wcu_stator+Wcurotor+Wnl;
efficiency_percent=kw*100/(kw+losses_total*0.001);
pf_noload=Icph/I0;
Rtotal_wrts=R_stator+R_rotor_wrts;

% leakage reactance
w0s=0.002;
h4=0.001;
h3=0.002;
h2=0.0005;
h1=dss-(h4+h3+h2); %meters
mu0=4*3.1416/10000000;
lemdass=mu0*(h1/(3*wss)+h2/wss+2*h3/(wss+w0s)+h4/w0s);
```

```
w0r=0.001;
h4=0.001;
h3=0.001;
h2=0.0005;
h1=dsr*0.001-(h4+h3+h2); %meters
mu0=4*3.1416/10000000;
wsr=wsr/1000; % conversion into meters
lemdasr=mu0*(h1/(3*wsr)+h2/wsr+2*h3/(wsr+w0r)+h4/w0r);
kwr=1;
lemdasr_wrts=lemdasr*(kws*kws)*Ss/(kwr*kwr*Sr);
lemdas=lemdass+lemdasr_wrts;
Xslot=8*3.1416*f*Ts*Ts*L*lemdas/(p*q);


% overhang leakage reactance
kwsleak=1;
lemda0=mu0*kwsleak*toy*toy/(3.1416*yss*L);
Xoverhang=8*3.1416*f*Ts*Ts*L*lemda0/(p*q);


% zigzag leakage reactance
qr=(Sr/p)/m;
Xm=vph/Imph;
Xzigzag=(5/6)*(Xm/(m*m))*(1/(q*q)+ 1/(qr*qr));
Xtotal_leak=Xoverhang+Xzigzag+Xslot; %total
Ztotal_leak=sqrt(Rtotal_wrts.^2+Xtotal_leak.^2);
Iscph=vph/Ztotal_leak;
pf_sc=Rtotal_wrts/ Ztotal_leak;


%% calculation of temp rise
dissip_stator=(2*L/Lmts)*Wcu_stator+(Wiron_ts+Wiron_cs);
surf_stator_out=3.1416*(D+2*dss+2*dcs)*L;
dissip_back_core=surf_stator_out/0.03; %watts/C
surf_stator_inn=3.1416*D*L;
speed_peri=3.1416*D*ns;
coeff1=0.04/(1+0.1*speed_peri);
dissip_inn_core=surf_stator_inn/coeff1; %watts/C
surf_ends=(2*3.1416/4)*((D+2*dss+2*dcs)^2 - (D^2));
air_velo=0.1*speed_peri;
coeff2=0.15/air_velo;
dissip_end_surf=surf_ends/coeff2; %watts/C
total_dissip=dissip_back_core+dissip_inn_core+dissip_end_surf; %watts/C
temperature_rise=dissip_stator/total_dissip;


%%% Design Sheet %%%%%%
disp({'dia of stator bore=' D ' meter'});
disp({'Length of stator= ' L ' meter'});
disp({'no load current= ' I0 ' Ampere'})
disp({'no load current= ' I0percent ' %'})
```

```
disp({'s.c. current= ' Iscph ' Amp'})
disp({'leak. imp./ ph= ' ztotal_leak ' ohm'})
disp({'resist. wrt stator=' Rtotal_wrts ' ohm'})
disp({'perc. efficiency= ' efficiency_percent ' %'})
disp({'slip= ' slip ' pu'})
disp({'rated current/ph= ' Iph ' Amp'})
disp({'n.l. power factor =' pf_noload})
disp({'s.c. power factor =' pf_sc})
disp({'temperature rise= ' temperature_rise ' degree centigrade'});
```

2.9.8 Review Exercises

I. Answer in Brief :

1. Write three methods to improve commutation in dc machines.
2. Write the conditions for
    (a) maximum power in induction motor
    (b) maximum torque in induction motor
    (c) maximum efficiency in induction motor.
3. Mention four disadvantages of harmonics in induction motors.
4. How can an induction motor be made to have a linear torque - slip characteristics in the slip rang 0–1?
5. Why do induction motor runs at low power factor when lightly loaded?
6. Is the resultant air gap flux / pole remains almost constant over the normal operating range of an induction motor. Justify your answer.
7. How can the direction of the stator revolving field be reversed?
8. What is the synchronous speed of 50 Hz induction motor that has an eight pole stator winding?
9. What is the maximum speed of magnetic field of a three phase 50 Hz induction motor can rotate?
10. Calculate the slip speed for question 8.
11. What is the rotor frequency when the rotor is stationary and stator is excited?
12. The rotor frequency is directly proportional to slip/synchronous speed/actual speed.
13. What is the material used for the rotor bars of induction motor?
14. What will be the efficiency and torque of induction motor under blocked rotor condition?
15. The speed of induction motor is always less than the synchronous speed. Why?
16. The cage bars of the squirrel cage induction motor are skewed. Why?
17. Why induction motors are widely used in industries? Give three reasons.
18. What happened to the motor power factor, if speed of motor is doubled at the same torque?
19. Write three factors which can affect the performance characteristics of induction motor.
20. Can we use pole changing method of speed control for slip ring induction motor? Justify your answer.

II. Multiple Choice Questions :

1. In an induction motor the stator mmf comprises
    (a) mmf equal to rotor mmf
    (b) mmf required to cancel rotor mmf
    (c) vector sum of magnetizing mmf and component to cancel rotor mmf vector
    (d) magnetizing mmf only.

2. The torque in an induction motor is given by the expression (ws= synchronous speed in mechanical rad(s):

(a) $(3I_2^2 R_2 /s)/(p*ws/2)$ (b) $3I_2^2 R_2 /s$

(c) $3I_2^2 R_2 / w_s)$ (d) $(3I_2^2 R_2 /s)/(p*ws)$.

3. At low slip the torque is

(a) inversely proportional to the square of the speed

(b) directly proportional to the square of the speed

(c) inversely proportional to the speed

(d) directly proportional to the speed

4. During short circuit test on induction motor (slip ring):

(a) the rotor is short circuited but is free to rotate

(b) the rotor is open circuited but not free to rotate

(c) the rotor is short circuited and also not free to rotate

(d) the rotor is open circuited and free to rotate.

5. For controlling the speed of an induction motor the frequency of supply is increased by 10%. For magnetizing current to remain the same, the supply voltage must:

(a) be reduced by 10%   (b) remain constant

(c) be increased by 10%   (d) be reduced or increased by 20%.

6. The shape of the torque/slip characteristic of the induction motor is

(a) hyperbola   (b) parabola

(c) straight line   (d) rectangular hyperbola.

7. If an induction motor has a slip of 2% at normal voltage. What will be the approximate slip when developing the same torque at 10% above normal voltage

(a) 1.6%   (b) 2%   (c) 1.65%   (d) 1.1%.

8. When the frequency of the induction motor is small, it can be measured by

(a) galvanometer   (b) dc moving coil mili-voltmeter

(c) dc moving coil ammeter   (d) dc voltmeter

9. The rotor efficiency of induction is called

(a) I.H.P.   (b) F.H.P.   (c) B.H.P.   (d) none.

10. Percentage tapping required of an auto-transformer for a cage motor to start the motor against one fourth of full load will be

(a) 70%   (b) 71%   (c) 71.5%   (d) 72.2%.

When the short circuit current on normal voltage is four times the full load and FL slip is 30%.

11. The value of the transformation ratio can be found by

(a) SC test only   (b) OC test only   (c) slip test   (d) stator resistance test.

12. The star-delta switch is equivalent to auto transformer of ratio (When applied to delta - connected cage induction motor)

(a) 57%   (b) 56.7%   (c) 86.6%   (d) 58% appx.

13. The power factor improvement of ac series motor is only possible by

(a) increasing the magnitude of inductances of field and armature windings.

(b) decreasing the magnitude of inductances of field and armature windings.

(c) equalising the armature resistance to armature reactance.

(d) none of the above.

14. Which of the following motor is unexcited single phase synchronous motor

(a) A.C. series motor   (b) universal motor

(c) repulsion motor   (d) reluctance motor.

15. The torque - slip characteristic of induction motor is shown in Figure 2.140, which is stable region

    (*a*) A   (*b*) B   (*c*) C   (*d*) D.

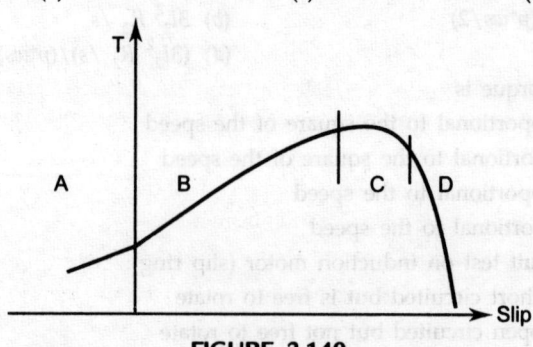

FIGURE 2.140

16. Magnetic induction results in

    (*a*) induced poles opposite from the original field poles.

    (*b*) induced poles same as the original field poles.

    (*c*) two north poles

    (*d*) An induced magnetic field but no poles.

17. The armature winding of repulsion motor is excited

    (*a*) conductively   (*b*) inductively   (*c*) resistively   (*d*) none.

18. The armature winding of series motor is excited

    (*a*) conductively   (*b*) inductively   (*c*) resistively   (*d*) none.

19. The large number of slots in induction motor

    (*a*) provide a better overload capacity

    (*b*) reduce overload capacity

    (*c*) provide bigger size of motor

    (*d*) reduce the size.

20. The good power factor of induction motor can be achieved if the average flux density in air gap is

    (*a*) large   (*b*) small   (*c*) infinite   (*d*) zero.

21. For a constant load torque the addition of rotor resistance in a slip ring induction motor

    (*a*) reduce stator current as well as slip

    (*b*) reduce stator current but increase its slip

    (*c*) cause no change in stator current but increase its slip

    (*d*) cause no change in stator current but decrease its slip.

22. As the load on induction motor increases

    (*a*) Its power factor goes on decreasing

    (*b*) Its power factor remains unchanged

    (*c*) Its power factor goes on increasing

    (*d*) Its power factor goes on increasing upto full load then it falls again.

23. A 400 kW, 3 phase, 440 V, 50 Hz, AC induction motor has a speed of 950 rpm on full load. The slip of machine will be

    (*a*) 0.06   (*b*) 0.04   (*c*) 0.01   (*d*) 0.05.

24. If the rotor of induction motor is assumed non-inductive the torque active on each conductor will be positive or unidirectional.

    (a) True   (b) False.

25. Rotor impedance seen from the stator is

    (a) $r_2' + js\ x_2'$ (b) $r_2 + js\ x_2$ (c) $r_2/s + js\ x_2$ (d) $r_2'/s + j\ x_2'$.

26. The resistance representing mechanical output in the equivalent circuit of an induction motor as seen from the stator side is

    (a) $r_2'(1/s - 1)$ (b) r_2'/s (c) $r_2^2(1/s - 1)$ (d) r_2/s.

27. At high slips the torque in an induction motor -

    (a) $T \propto (1/s)$ (b) $T \propto s^2$ (c) $T \propto (1/s^2)$ (d) $T \propto s$.

28. The shunt resistance component in equivalent circuit obtained by no load test of an induction motor is representative of

    (a) windage and frictional losses only   (b) core losses only
    (c) Core, windage and frictional losses   (d) copper losses.

29. By adding a resistance in the rotor circuit of a slip ring induction motor

    (a) the starting current and torque both reduced (compared to DOL)
    (b) the starting current and torque both increased
    (c) the starting current reduced and torque increased
    (d) the starting current increased and torque reduced

30. The power input to an induction motor is 40 kW when it is running at 5 % slip. The stator resistance and core losses are assumed negligible. The torque developed in synchronous watts is

    (a) 42 kW   (b) 40 kW   (c) 38 kW   (d) 2 kW.

31. A squirrel cage induction motor having a rated slip of 2 % on full load has a starting torque of 0.5 full load torque. The starting current

    (a) equal to full load current   (b) twice of full load current
    (c) four times the full load current   (d) five times the FL current.

32. Make the correct answer as the load of an induction motor is increased upto full load :

	PF	Slip	Efficiency
(a)	increases	increases	increases
(b)	decreases	increases	increases
(c)	decreases	decreases	increases
(d)	decreases	decreases	decreases

33. An eight pole single phase induction motor is running at 690 rpm. Its slips with respect to the two fields is

	Forward field	Backward field
(a)	0.08	1.92
(b)	1.92	0.08
(c)	0.08	2.00
(d)	1.00	2.00

34. In a slip ring induction motor resistance is connected in rotor phases

    (a) To limit starting current
    (b) To increase starting torque
    (c) To limit starting current and to increase starting torque
    (d) None of the above.

35. An induction motor is
    (a) Self starting with very high torque
    (b) Self starting with very low torque
    (c) Self starting with low torque
    (d) Self starting with high torque

36. The crawling in induction motor is caused by
    (a) Improper design of machine   (b) low voltage supply
    (c) High loads   (d) Harmonics developed in the motor.

37. If a three phase motor is operated on single phase it will
    (a) Burn out   (b) run at rated speed
    (c) run with lot of noise   (d) run smoothly.

38. Slip rings of induction motor are made of
    (a) Aluminum   (b) Carbon
    (c) Phosphor Bronze   (d) Cobalt steel.

III. Fill in the Blanks :

1. The difference between the synchronous speed and the actual speed of an induction motor is known as ....

2. Slip ring of an induction motor is made of ....

3. A .... induction motor is more sensitive to fluctuation in supply voltage.

4. In a wound rotor induction motor the rotor windings are shorted through ....

5. In a wound rotor induction motor the 3 phase windings are .... connected.

6. The slip of induction motor at no-load will be .... and at blocked rotor condition ....

7. In double cage induction motor
    (a) Inner cage has .... resistance and .... reactance.
    (b) Outer cage has .... resistance and .... reactance.

8. Speed of differential cascade connected set is .... cumulative cascade connected set.

9. The maximum speed of magnetic field of a three phase 50 Hz induction motor is ....

10. The rotor frequency will be .... the stator when rotor is stationary and stator is excited.

Answers

II. Multiple Choice Questions :

1. (c)	**2.** (d)	**3.** (c)	**4.** (c)	**5.** (c)
6. (d)	**7.** (c)	**8.** (b)	**9.** (b)	**10.** (d)
11. (a)	**12.** (d)	**13.** (b)	**14.** (d)	**15.** (d)
16. (a)	**17.** (b)	**18.** (a)	**19.** (b)	**20.** (b)
21. (b)	**22.** (d)	**23.** (d)	**24.** (a)	**25.** (d)
26. (a)	**27.** (a)	**28.** (c)	**29.** (c)	**30.** (b)
31. (d)	**32.** (a)	**33.** (a)	**34.** (c)	**35.** (c)
36. (c)	**37.** (c)	**38.** (c)		

SECTION-F

2.10 SINGLE PHASE INDUCTION MOTOR

The electric motor converts electrical energy into mechanical energy to perform some physical task or work.

2.10.1 Introduction

In order to understand how the single-phase AC induction motor works a basic understanding of the physical principles and fundamentals governing motor and operation is required. The basic operation of an AC induction motor is based on two electromagnetic principles :

1. Current flow in a conductor will create a magnetic field surrounding the conductor, and,
2. If a conductor is moved through the magnetic field, current is induced in the conductor and it will create its own magnetic field.

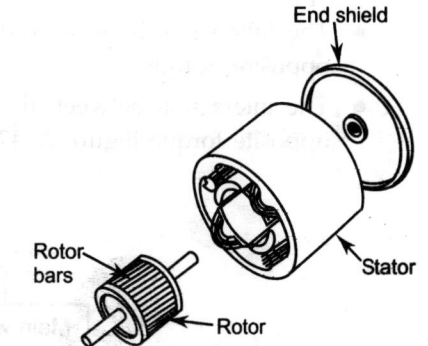

The fundamental single-phase AC induction motor consists of three basic parts; namely, stator, rotor and bearings (refer Figure 2.141). The stator of single phase induction motor consists of distributed or concentrated copper windings and rotor contains short circuited bars.

The single-phase induction machine is the most frequently used motor for refrigerators, washing machines, drills, compressors, pumps, and so forth.

The stator of a single-phase induction motor shown in Figure 2.142 has a laminated iron core with two windings arranged perpendicularly.

FIGURE 2.141. *Basic components of a single-phase AC induction motor*

- One is the main and
- The other is the auxiliary winding or *starting winding*

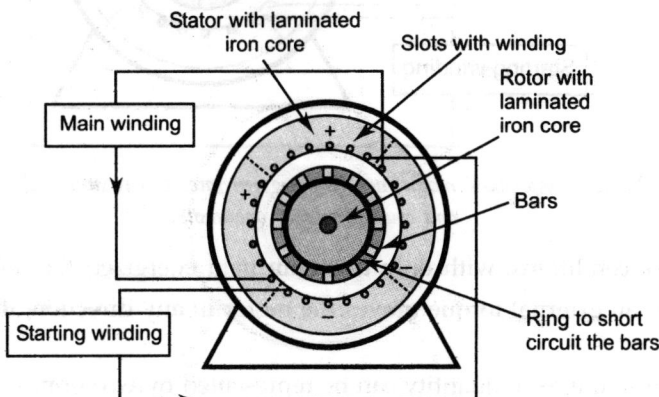

FIGURE 2.142. *Single-phase induction motor*

Stator. The stator is constructed of a set of stacked laminated discs which are surrounded by a stator winding. This winding is connected to AC supply which produces a magnetic field that **revolves** at a speed called "synchronous speed."

Rotor. The rotor is connected to the output shaft and consists of short circuited bars which is casted into slots and joined at both ends with end rings. The rotor when placed in the stator magnetic field creates a magnetic field of its own and interacts with the magnetic field of the stator and produce torque.

- The motor uses a squirrel cage rotor, which has a laminated iron core with slots.
- The rotor bars are molded on the slots and short-circuited at both ends with a ring.

2.10.2 Operating Principle

The single-phase induction motor operation can be described by double revolving field theory, which may be explained as:

- A single-phase ac current supply to the main winding that produces a pulsating magnetic field.
- Mathematically, the pulsating field could be divided into two fields, which are rotating in opposite directions.
- The interaction between the fields and the current induced in the rotor bars generates opposing torque.
- The interaction between the fields and the current induced in the rotor bars generates opposite torque Figure 2.143.

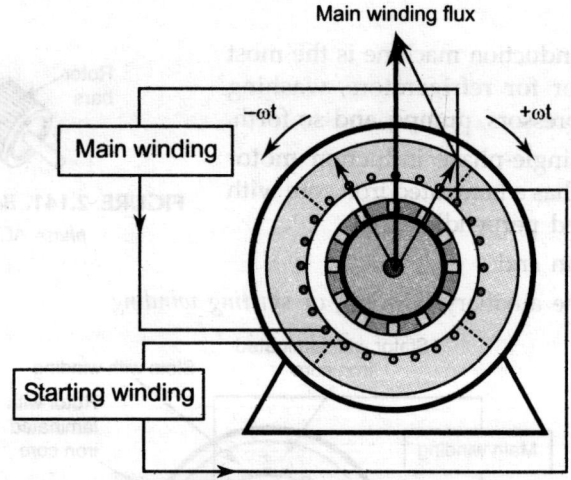

FIGURE 2.143. *Single-phase motor main winding generates two rotating fields, which oppose and counter-balance one another*

- Under these conditions, with only the main field energized the motor will not start.
- However, if an external torque moves the motor in any direction, the motor will begin to rotate.
- An alternating uni-axial quantity can be represented by two oppositely-rotating vectors of half magnitude. Accordingly, an *alternating* sinusoidal flux can be represented by two revolving fluxes, each equal to half the value of the alternating flux and each rotating synchronously ($Ns = 120 \ f/P$) in the opposite direction.

For example, a flux given by $\Phi = \Phi_m \cos 2\pi ft$ is equivalent to two fluxes revolving in opposite directions, each with a magnitude of $\Phi/2$ and an angular velocity of $2\ \pi f$. Eulers expression for $\cos\theta$ provides interesting justification for the decomposition of a pulsating flux.

Eulers expression is $\cos\theta = \dfrac{e^{j\theta} + e^{-j\theta}}{2}$.

The term $e^{j\theta}$ represents a vector rotated clockwise through an angle θ whereas $e^{-j\theta}$ represents rotation in anticlockwise direction. Now the above given flux can be expressed as $\phi_m \cos(2\pi ft)$

$= \dfrac{\phi_m}{2}\left(e^{j2\pi ft} + e^{-j2\pi ft}\right)$. The right-hand expression represents two oppositely-rotating vectors of half magnitude as shown in Figure 2.144.

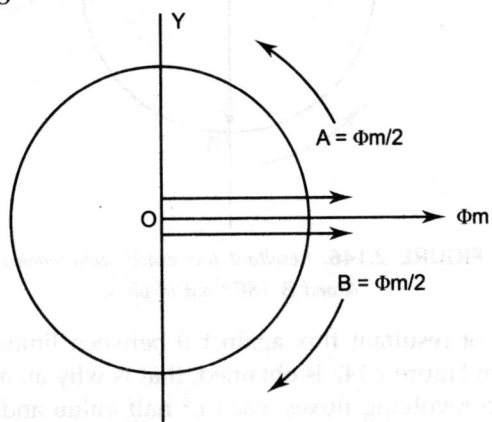

FIGURE 2.144. *Two oppositely-rotating flux vectors*

Let the alternating flux have a maximum value of Φ_m. Its component fluxes A and B will each be equal to $\Phi_m/2$ revolving in anticlockwise and clockwise directions respectively. After some time, when A and B would have rotated through angle $+\theta$ and $-\theta$, as in Figure 2.145, the resultant

flux would be $2x \dfrac{\phi_m}{2} \cos\dfrac{2\theta}{2} = \phi_m \cos\theta$.

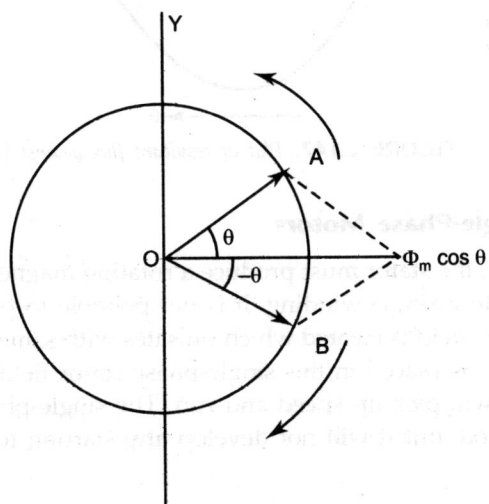

FIGURE 2.145. *Resultant flux*

After a quarter cycle of rotation, flux *A* and *B* will be oppositely-directed as shown in Figure 2.146 below so that the resultant flux would be zero.

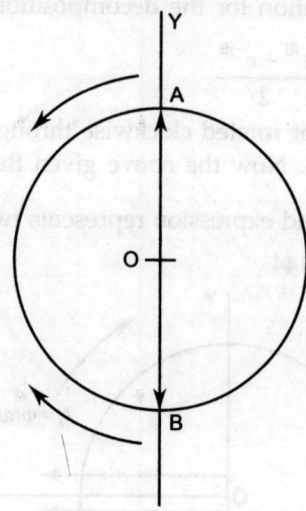

FIGURE 2.146. *Resultant flux exactly zero when flux A and B 180° out of phase*

If we plot the values of resultant flux against θ between limits $\theta = 0$ to 360, then a curve similar to the one shown in Figure 2.147 is obtained, that is why an alternating flux can be looked upon as composed of two revolving fluxes, each of half value and revolving synchronously in opposite direction.

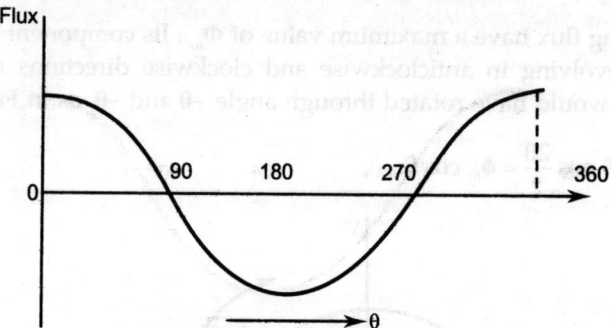

FIGURE 2.147. *Plot of resultant flux against θ*

Magnetic Field for Single-Phase Motors

In order to rotate the rotor, the stator must produce a rotating magnetic field. With a single source of AC voltage connected to a single winding, it is not possible to produce a revolving magnetic field. A stationary magnetic field is created which pulsates with same frequency as the AC voltage varies. If a stationary rotor is placed in this single-phase stator field, it will not rotate, but if the rotor is spun by hand, it will pick up speed and run. The single-phase motor will run (in either direction) if started by hand, but it will not develop any starting torque as illustrated in Figure 2.148.

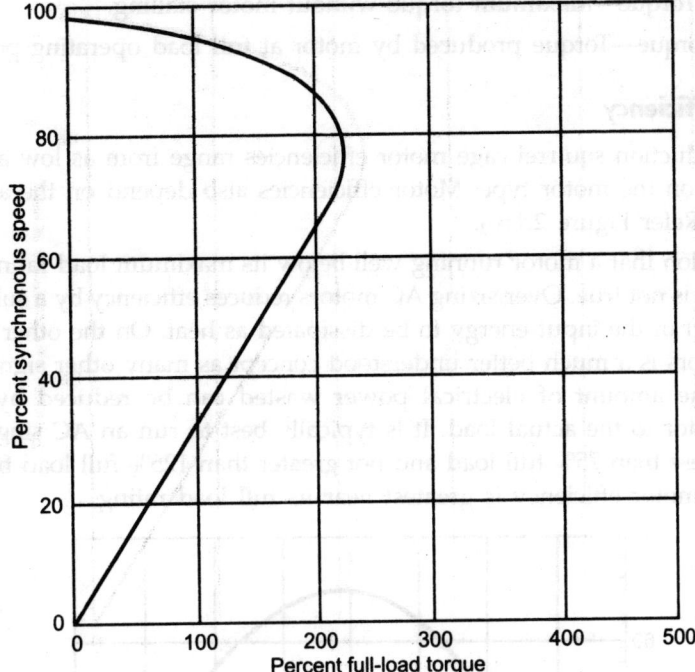

FIGURE 2.148. *Speed-Torque Characteristics of Single-Phase with main winding only*

2.10.3 Motor Speed-Torque Curves

There is much information on a speed-torque curve to tell the end user if the motor will operate satisfactorily for the intended application. The speed-torque curve will allow the user to determine if the motor has enough starting torque to overcome friction, to accelerate the load to full running speed, and if it can handle the maximum overload expected (Ref. Figure 2.149).

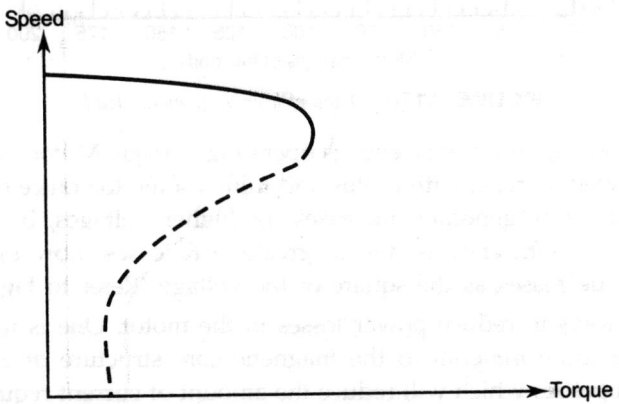

FIGURE 2.149. *Typical motor speed-torque curve*

There are many torque that can be obtained from a motor's speed-torque curve:

- Starting Torque—Motor torque at zero speed.
- Pull-Up Torque—Lowest torque value between zero and full load speed.

- Breakdown Torque—Maximum torque without motor stalling.
- Full Load Torque—Torque produced by motor at full load operating point.

2.10.4 Motor Efficiency

Single-phase AC induction squirrel cage motor efficiencies range from as low as 30% to as high as 65%, depending on the motor type. Motor efficiencies also depend on the actual motor load versus rated load (Refer Figure 2.150).

The misconception that a motor running well below its maximum load rating will run cooler and more efficiently is not true. Over sizing AC motors reduces efficiency by a substantial amount, causing a larger part of the input energy to be dissipated as heat. On the other end of the scale, overloading of motors is a much better understood concept as many other signs indicate a poor motor selection. The amount of electrical power wasted can be reduced by a more careful application of a motor to the actual load. It is typically best to run an AC single-phase squirrel cage motor at not less than 75% full load and not greater than 125% full load from an efficiency standpoint. Again, motor efficiency is greatest near its full load rating.

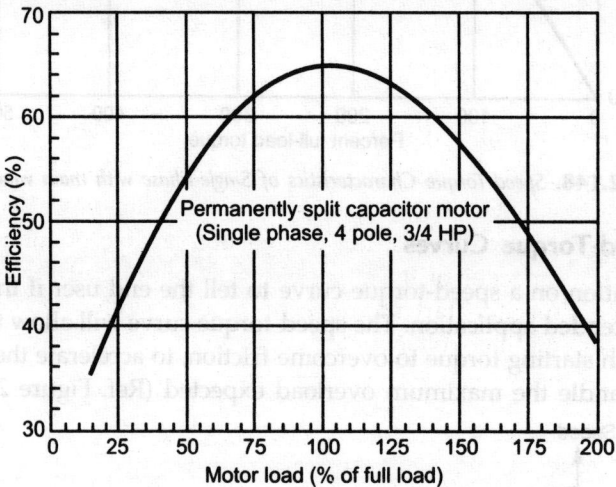

FIGURE 2.150. *Motor efficiency vs. motor load*

Another criteria affecting motor efficiency is operating voltage. Motors are generally designed to operate at a given rated voltage, with a plus and minus some tolerance (typical ±10%). Within the tolerance level, efficiency generally increases for higher voltages, but decreases for lower voltages. The decrease in efficiency is due to greater I^2R losses. Low operating voltage also reduces torque, which decreases as the square of the voltage (Refer to Figure 2.151).

There are several ways to reduce power losses in the motor. One is to reduce losses in the core, either by adding more material to the magnetic core structure or by using a steel with improved core-loss properties which will reduce the amount of current required to magnetize the core. Another method is to increase the cross-sectional area of conductors to reduce resistance. This means that additional winding material must be added to the stator and rotor. Another alternative is to shorten the air gap between the rotor and stator to reduce the magnetizing current required.

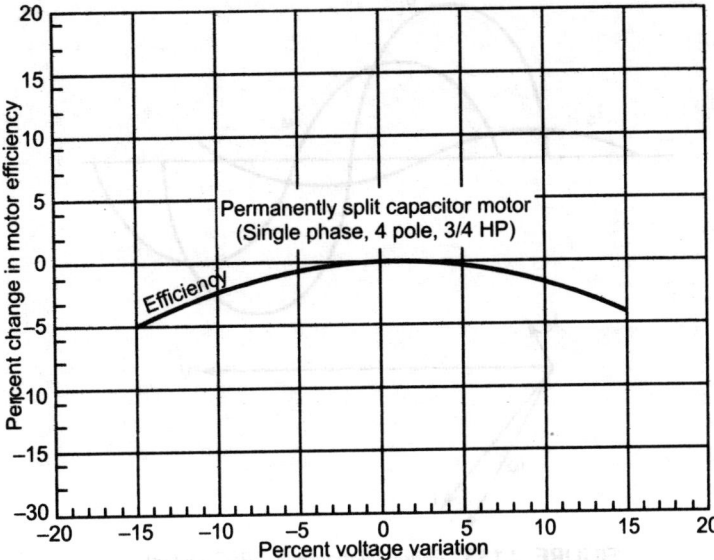

FIGURE 2.151. *Effect of voltage on motor efficiency*

Permanently Split Capacitor Motors (PSC)

Figure 2.152 illustrates schematically the winding arrangement of a typical distributed winding of a permanently split capacitor motor. The windings of the PSC motor are arranged like those of the split-phase and capacitor start designs, but a capacitor capable of running continuously replaces the intermittent duty capacitor of the capacitor- start motor and the centrifugal switch of both the split phase and capacitor-start motors. The main winding remains similar to the previous designs, current lags the line voltage (refer to Figure 2.152).

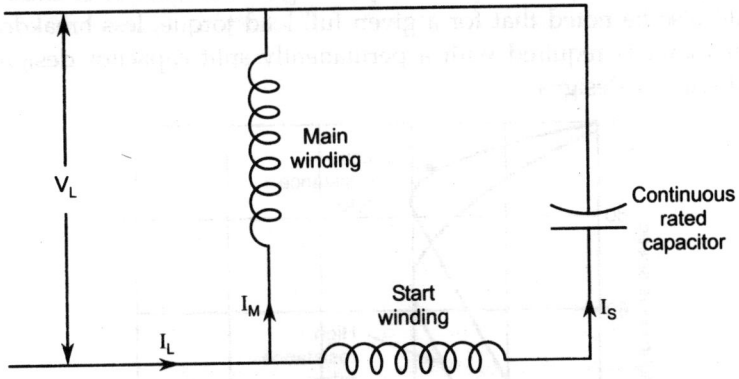

FIGURE 2.152. *Winding schematic for PSC motor*

The start winding of a PSC motor is somewhat different than in the capacitor-start design. Because the capacitor for a PSC motor usually has a small rating, it is necessary to boost the capacitor voltage by adding considerably more turns to its coils than are in the main winding coils. Start winding wire size remains somewhat smaller than that of the main winding. The smaller microfarad rating of the capacitor produces more of a leading phase shift and less total start winding current as shown in Figure 2.153, so starting torques will be considerably lower than with the capacitor-start design.

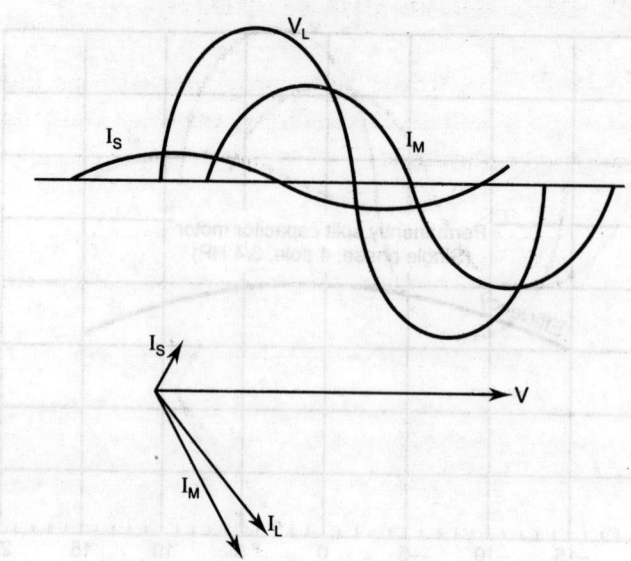

FIGURE 2.153. *Phase relationships (PSC motor)*

However, the real strength of the permanently split capacitor design is derived from the fact that the start winding and capacitor remain in the circuit at all times and produce an approximation of two-phase operation at the rated load point. This results in better efficiency, better power factor, and lower 100 Hz torque pulsations than in equivalent capacitor-start and split-phase designs.

Figure 2.154 illustrates a typical speed torque curve for a permanently split capacitor motor. Different starting and running characteristics can be achieved by varying the rotor resistance. In addition, by adding extra main windings in series with the original main windings, PSC motors can be designed to operate at different speeds depending on the number of extra main windings energized. It should also be noted that for a given full load torque, less breakdown torque and therefore a smaller motor is required with a permanently split capacitor design than with the other previously discussed designs.

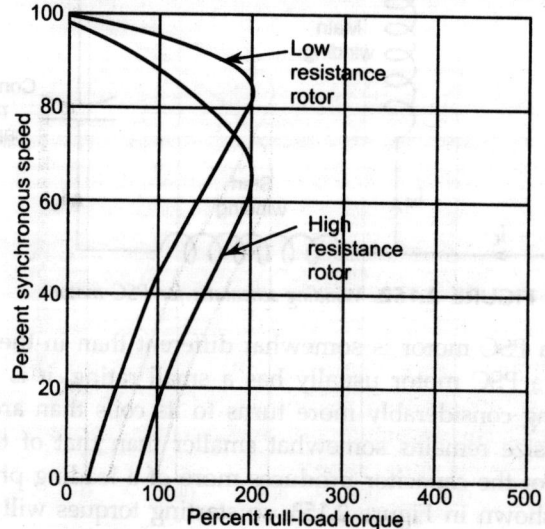

FIGURE 2.154. *General performance characteristic*

2.11 EXPERIMENT ON SINGLE PHASE MOTOR

2.11.1 No Load and Blocked — Rotor Tests on a Single-phase Induction Motor (capacitor start type) and to Derive form the Test Data its Equivalent Circuit for Running Conditions

Theory

The equivalent circuit of a single-phase induction motor (based on the revolving field theory) with only main winding (also called running winding) effective, is shown in Figure 2.155.

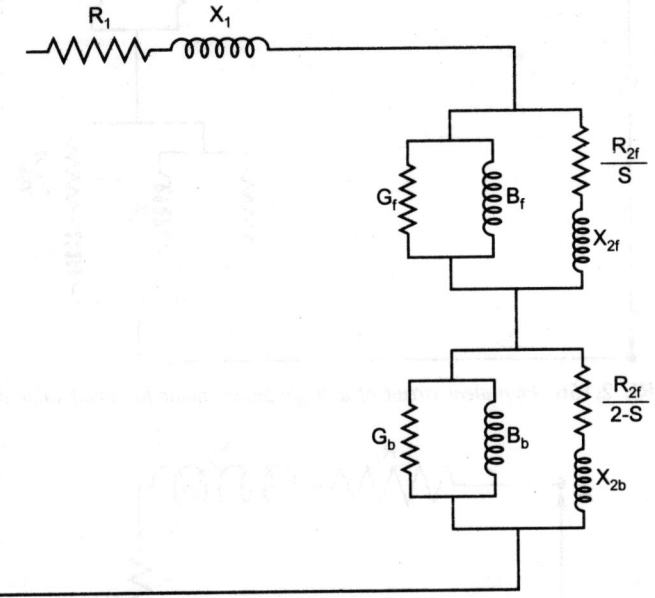

FIGURE 2.155. *Equivalent circuit of a single-phase motor*

where

R_1 = Main winding effective resistance,

X_1 = Main winding leakage reactance,

Z_f Rotor impedance with respect to the forward rotating field, referred to stator,

Z_b = Rotor impedance with respect to the to the backward rotating field, referred to stator, = $R_{2b} + jX_{2b}$

Y_t = Admittance representing excitation characteristics for the backward filed. = $G_t - jB_t$

Y_b = Admittance representing excitation characteristics for the backward field= $G_b - jB_b$

S = Fractional slip with respect to the forward field.

Under no-load conditions, is very small, R_{21}/s becomes very large, and the equivalent circuit reduces to that shown in Figure 2.156. At standstills is unity and the equivalent circuit takes, the form shown in Figure 2.157. The parameters of the equivalent circuit can be calculated from the data obtained from the following three tests.

1. **No-load test (Running-light load test):** With the motor running light load at a rated voltage, rated frequency source, the input current I and the input power p are recorded. The auxiliary winding must not be in circuit while recording the observations.

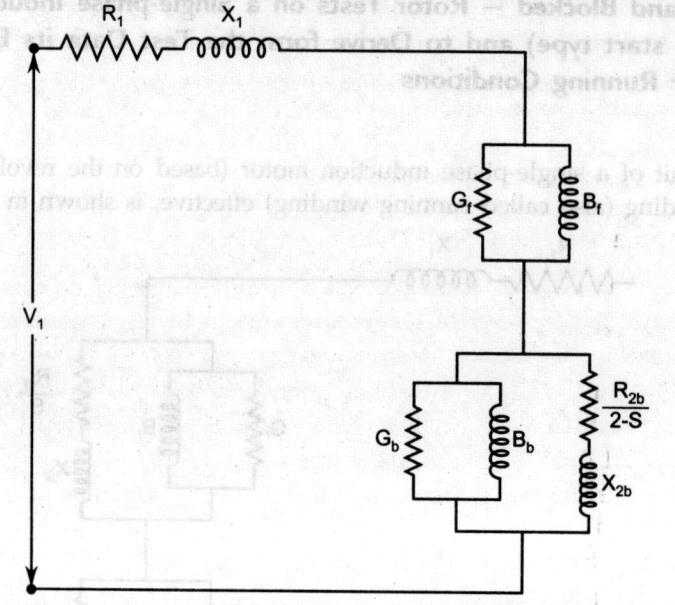

FIGURE 2.156. *Equivalent circuit of a single-phase motor for small value of slips*

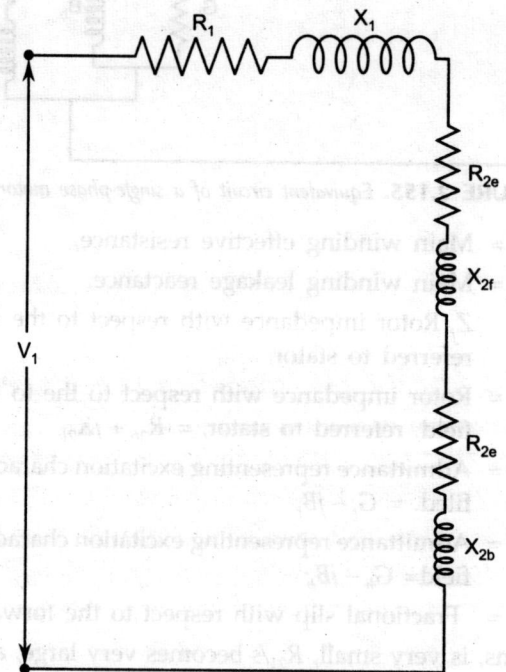

FIGURE 2.157. *Equivalent circuit of a single-phase motor under Blocked rotor conditions*

2. Blocked-rotor test with rotor blocked and a reduced voltage applied to the main winding, the input current I_{sc}, the input power P_{sc} and the voltage applied V_{sc} are recorded. The auxiliary winding should not be in circuit during this test as well.

3. **Measurement of d.c. resistance of main winding:** This may be measured by the voltmeter-ammeter method, immediately after the blocked rotor test, to get the value at actual winding temperature. Alternatively, the resistance may be measured with the winding at room temperature and hot resistance computed from that value. This hot resistance of the main winding is hereafter denoted by R_1.

Computation of Parameters from Test Data

(a) **From blocked rotor test:** If Z_{sc} is the equivalent impedance of the machine during blocked rotor test. From Z_{sc} and power factor under short circuit test these relations R_{sc} and X_{sc} may be obtained.

As the blocked-rotor test is performed at a reduced voltage, the exciting current and core loss during this test are very small and may be neglected. In terms of the equivalent circuit of Figure 2.157, assuming excitation branch admittances to zero. Further, at $s = 1$, the frequency of the currents on account of both the forward field and the backward field are same. Under this condition, $R_{2e} = R_{2b} = R_{2f}$. With these two approximations the equivalent circuit of Figure 2.155 may be redrawn as shown in Figure 2.157.

If X_2 is the total rotor reactance (= $X_{2f} + X_{2b}$), then $X_1 + X_2 = X_{sc}$. It is very difficult to separate X_1, X_2 but is customary to consider $X_1 \approx X_2$.

Therefore, $X_1 \approx X_2 \approx \dfrac{1}{2}$. Under blocked-rotor conditions, it is easy to visualize that

$$X_{2f} = X_{2b}$$

$$\therefore \qquad X_{2f} = X_{2b} = \tfrac{1}{2} X_2$$

The effective value of main winding resistance at line frequency is usually 1.1 to 1.3 times the d.c. value, the actual ratio depending upon conductor configuration etc. Thus R_1. It is seen from Figure 2.157 that

$$R_{sc} = R_1 + 2R_{2e}$$

Therefore, $R_{sc} = \tfrac{1}{2} (R_{sc} - R_1)$

R_{2e} being the effective rotor resistance at line frequency.

Under normal running conditions, the forward field-slip is small and, therefore, the rotor resistance for this field will be very nearly the d.c. resistance, *i.e.*, R_2 divided by a suitable factor to reduce it to its d.c. value. This factor has a value in the range.

Therefore $R_{2dc} = R_{2f} = R_{2e} /(\text{factor } 1.2 - 1.4)$.

With respect to the backward field the frequency of the rotor currents under actual running conditions is approximately twice the line frequency. Effective rotor resistance, with respect to the backward field at this high frequency, is generally $1.6 - 1.8 \, R_{2dc}$. Therefore $R_{2b} = 1.6 - 1.8 \, R_{2dc}$.

Thus, from known values of R_1, blocked-rotor test data and suitable choice of factors for converting effective resistance into d.c resistances and vice versa, R_{2f}, R_{2b}, X_{2f}, X_{2b} and X_1 are obtained.

(b) **From no-load test:** As pointed out earlier, under no-load conditions the impedance of the load branch in the equivalent circuit for forward field is very high (slip small) and may be considered infinite. Further, the voltage across backward field load branch under no-load conditions is so low that the exciting branch for the backward field may be neglected, *i.e.*, the admittance of the exciting branch for backward field is assumed to be zero. The equivalent circuits of Figures 2.156 and 2.157 then reduce to the form shown in Figure 2.158.

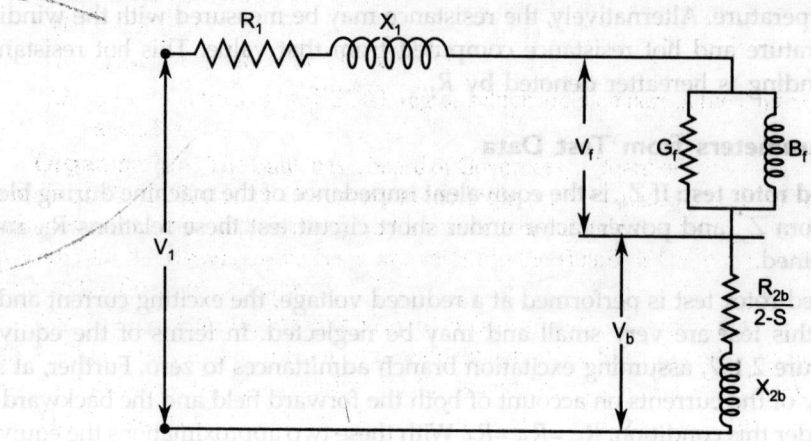

FIGURE 2.158. *Simplified equivalent circuit under no-load test*

The total no-load power input is used up in stator copper loss ($I_0^2 R_1$), backward-field resistance copper loss $\left(I_0^2 \dfrac{R_{2b}}{2} \right)$, forward field excitation (iron) loss and backward field excitation loss. Under no load conditions, the magnitude of the backward rotating field is very small and, therefore, the backward field iron-loss may be neglected.

Forward field iron loss $P_i = P_o - I_0^2 R_1 - I_0^2 R_2 / 2$.

Voltage across the forward exciting circuit

$$= V_1 - I_0 \left(R_1 + jX_1 - jX_{2b} + \frac{2_{2b}}{2} \right)$$

Exciting circuit admittance $= \dfrac{I_0}{V_f}$

Exciting circuit conductance $= \dfrac{P_{if}}{V_f^2}$

Thus, all the parameters of the equivalent circuit pertaining to running conditions are known.

An alternative approximate form of equivalent circuit is given in Figure 2.158 and the method of calculating its parameters.

Experimental Details

Capacitor-start single-phase induction motors are usually provided with a centrifugal switch placed on the rotor shaft and connected in series with the auxiliary winding (also called stating winding) circuit. When the motor is at rest, the switch is closed and the starting winding is in circuit, and the motor, therefore be started as a split-phase motor. When the shaft has attained a speed equal to about 80% of the rated speed, the centrifugal switch opens, thereby disconnecting the starting winding. While performing the blocked-rotor test, the switch is normally closed, whereas the test conditions require that the auxiliary winding circuit should be open. An additional switch must, therefore, be provided in the starting winding circuit.

For blocked-rotor test, the voltmeter should have a range equal to about 20% of the rated voltage of the motor, whereas the ammeter should have a range about twice full-load current of the machine. For running-light test, the voltmeter should be of a range equal to about 150% of rated voltage of machine, and the ammeter should have a range equal to half the full-load current of the machine.

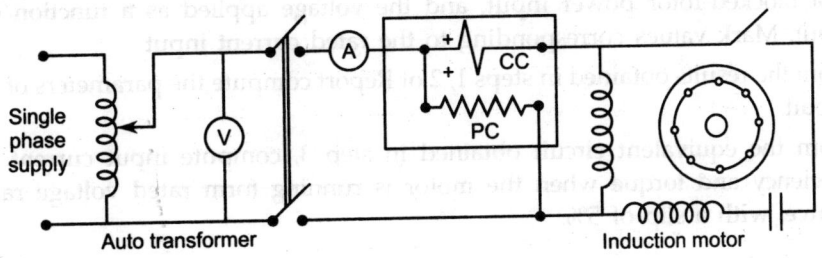

FIGURE 2.159. *Connection diagram of no-load and blocked rotor tests*

Procedure

(a) **Running-light test:**

1. Connect for no-load test as shown in Figure 2.159.

2. Adjust the auto-transformer to such a value that its output voltage is equal to the rated voltage of the motor.

3. If switch S_2 is a centrifugal switch, it will be closed when the motor is at rest. Otherwise close S_2 and then close S_1 at the time of motor starting. When the motor has attained a steady speed, open the switch S_2, in case it is an external switch. Record V, I ad power input P.

4. Increase the applied voltage by 10% and record the current and power input, Reduce V in suitable steps, ad record the current and power input for several values of V.

5. Stop the motor.

(b) **Blocked-rotor test:**

1. Connections for blocked-rotor test are the same as those for no-load test, except that the instruments have to be replaced by those of suitable ratings.

2. Disconnect the starting winding.

3. Adjust the auto-transformer such that its output voltage is about 5% of the rated voltage of the motor.

4. Close switch S_1 and observe the current, power input and applied voltage.

5. Adjust V to such a vale that the input current is 50% of the rated current. Record applied voltage, input current and power.

6. Increase the applied voltage, and record data as in step 5, till the current input is equal to the rated current of the motor.

7. Switch off the motor. (c) Immediately after the blocked-rotor test, measure the resistance of the running winding using voltmeter-ammeter method.

Results

1. *(a)* Plot no-load current, power input as a function of applied voltage. Obtain therefrom the no-load current I and power input P corresponding to rated voltage of the motor.

   (b) Extrapolate the curve of P, up to zero-voltage to obtain the friction and windage loss.

2. Plot blocked-rotor power input, and the voltage applied as a function of the current input. Mark values corresponding to the rated current input.

3. From the results obtained in steps 1, 2 of Report compute the parameters of the equivalent circuit.

4. From the equivalent circuit obtained in step 3, compute input current, input power, efficiency and torque when the motor is running form rated voltage rated frequency source, with a slip of 5%.

Discussion

The equivalent circuit, derived in this experiment, is based on a number of approximations. R_2 and R_s are the effective values of resistances and, therefore, change with slip. For the normal range of running speeds, however, (slips 1 -5%) these effective values may be assumed to remain constant. The leakage reactance is assumed to be equally divided between the stator and rotor, whereas actually it may not be so. The factors for converting effective values of resistances into d.c. values and vice versa are arbitrary and cannot be easily determined.

The most important assumption in the experiment is that the friction and windage losses, which are mechanical, are combined with the other losses in calculating the loss conductaces G G_f, G_b. For a more accurate analysis, the friction and windage loss should be subtracted form the mechanical power developed, and should not be included along with excitation losses. To determine the friction and windage loss, no-load losses are plotted as a function of applied voltage. Extrapolation of this curve to zero applied voltage gives the friction and windage loss (Figure 2.160). The difference of the no-load loss at rated voltage and the friction and windage loss is the no-load copper and iron losses. This corrected loss should have been used for P in the calculations from no-load test data.

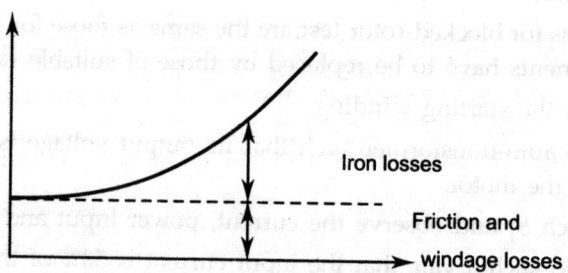

FIGURE 2.160. *Determination of friction and windage losses*

An alternative approximation, the core losses and friction and windage losses are lumped together and treated as a mechanical loss to be subtracted form the mechanical power developed. In this case, the core loss conductance G_f and G_b are reduced to zero, and the equivalent circuit simplified.

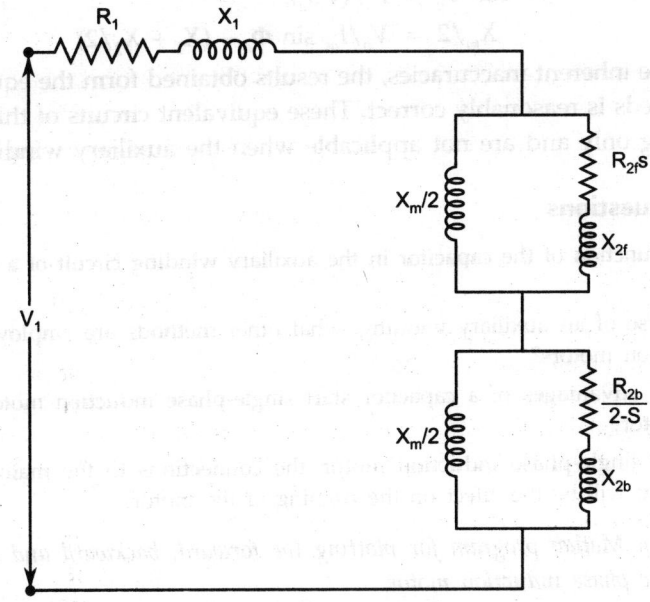

FIGURE 2.161. *Alternative equivalent circuit of a single-phase motor*

This equivalent circuit, with a slightly different notation is shown in Figure 2.161.

where R_2 – equivalent rotor resistance referred to stator

X_2 – rotor leakage reactance at standstill, referred to stator.

X_m – Magnetizing reactance, referred to stator.

The parameters of this equivalent circuit with certain approximations may be calculated in the following manner.

(a) **From blocked rotor test:** If Z_{sc} is the equivalent impedance of machine under blocked rotor test, V_{sc} is the voltage applied to main winding, and I_{sc} is the current in the main winding

$$Z_{sc} = V_{sc} / I_{sc}$$
$$R_{sc} = R_1 + R_2 = P_{sc} / (I_{sc}2)$$
$$X_{sc} = X_1 + X_2 = (Z_{sc}2 - R_{sc}2)$$

and $$X_1 = X_2 = X_{sc} / 2$$

On the assumption that $X_1 = X_2$. Further, R_1 should be the effective value of stator resistiance, i.e., $R_1 = k . R_{1dc}$ where k is a factor in the range 1.1 – 1.3. It is usually 1.1 for small motors.

(b) **From light load test:** If V_o, I_o and P_o are applied voltage, input current an input power respectively during the light load test, in which the speed is nearly synchronous speed and slip $s = 0$.

$$Z_0 = (R_1 + jX_1) + jX_m/2 + R_2/4 + jX_2/2.$$

This assumes that in the part of the equivalent circuits representing the backward field, $X_m/2 >> [(R_2/2 + j X_2/2)]$.

The magnetizing reactance branch is considered to be open.

Also,
$$\cos \Phi_o = P_o/(V_o I_o);$$
$$X_m/2 = V_o/I_o, \sin \Phi_o - (X_1 + X_2/2)$$

In spite of all the inherent inaccuracies, the results obtained form the equivalent circuit at the normal running speeds is reasonably correct. These equivalent circuits of this experiment pertain to the main winding only and are not applicable when the auxiliary winding is also in circuit.

2.11.2   Review Questions

1. What is the function of the capacitor in the auxiliary winding circuit of a single-phase induction motor?

2. Besides the use of an auxiliary winding, what other methods are employed for starting single-phase induction motors?

3. What are the advantages of a capacitor start single-phase induction motor over a shaded-pole induction motor?

4. In a running single-phase induction motor, the connections to the main winding are quickly reversed. What will be the effect on the running of the motor?

PROBLEM. *Write a Matlab program for plotting the forward, backward and total torque Vs speed characteristics of single phase induction motor.*

2.11.3   Matlab Program to Draw Speed–Torque Characteristics of Single Phase Induction Motor

```
%*********************************************************************
%
% Tw_1ph_im.m - Torque-speed characteristics for 1-phase induction motor
%
%*********************************************************************
clear all;
clf;
p=4; f=60; Vm=120*sqrt(2); % p=poles, f=frequency
% Equivalent circuit paramters of single phaase induction motor
Rs=2.02; Xs=2.79; Xm=106.8; Rr=4.12; Xr=2.12;
smax=2; npts=100; % smax=max slip, npts=points for curve
ws=2/p*2*pi*f; % Synchronous speed
Zs=Rs+Xs*j;
Zm=0+Xm/2*j;
s=zeros(1,npts); dels=smax/npts;
for i=2:npts-1;
   s(i)=(i-1)*dels;
end
s(1)=0.001; s(npts)=smax-0.001;
Td=zeros(1,npts); Tdf=Td; Tdb=Td; wm=Td;
for i=1:npts % Loop to calculate torque points
   ZU=Rr/2/s(i)+Xr/2*j;
   ZL=Rr/2/(2-s(i))+Xr/2*j;
```

```
    Zf=Zm*ZU/(Zm+ZU);  Zb=Zm*ZL/(Zm+ZL);
    Is=Vm/sqrt(2)/(Zs+Zf+Zb);
    Irf=abs(Is*Zm/(Zm+ZU));
    Irb=abs(Is*Zm/(Zm+ZL));
    wm(i)=(1-s(i))*ws;
if wm(i) == 0
    Tdf(i)=0.5*Irf^2*Rr/s(i)/ws;
    Tdb(i)=-0.5*Irb^2*Rr/(2-s(i))/ws;
else;
    Tdf(i)=0.5*Irf^2*Rr*(1-s(i))/wm(i)/s(i);
    Tdb(i)=0.5*Irb^2*Rr*(s(i)-1)/wm(i)/(2-s(i));
end
    Td(i)=Tdf(i)+Tdb(i);
    nm=wm*30/pi;
end
plot(nm,Td,nm,Tdf,'-',nm,Tdb,'-.');grid
title('Torque-speed for single-phase induction motor');
xlabel('Speed, rpm'); ylabel('Torque, N-m');
legend(' Total torque','Forward T ',' Backward T ',1)
%*************************************************************
```

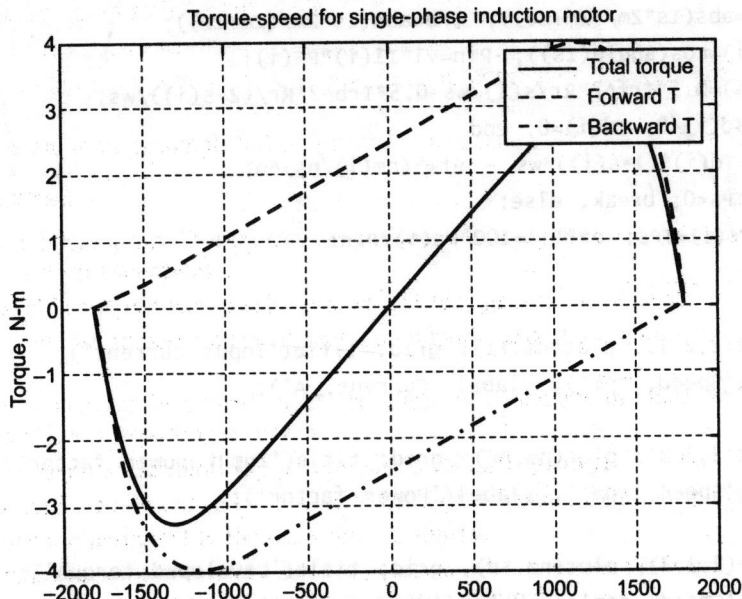

FIGURE 2.162. *Torque-speed characteristics for single phase induction motor.*

PROBLEM. *Write a Matlab program for simulating the performance (stator current, power factor, Torque and efficiency) of single phase induction motor from the equivalent circuit parameters. Assumes friction and windage losses vary as cube of speed.*

2.11.4 Matlab Program for Calculating Performance of s-phase Induction Motor

```
%*******************************************************************
% sp_perf.m - calculates single-phase induction motor performance
%*******************************************************************
clear all;
% Name plate Specifications
V1=120; f=50; p=4; % Phase voltage, frequency, poles
% Equivalent circuit parameters
Rs=2.02; Xs=2.79; Xm=106.8; Rr=4.12; Xr=2.12;

Pfw=10.5; n=3; % Total F&W losses at syn. speed, speed dependence
npts=100; s=linspace( 0.0001,1,npts); s=fliplr(s);
I1=zeros(1,npts); Td=I1; PF=I1; Ps=I1; eff=I1; nm=I1;
ws=2/p*2*pi*f; ns=120*f/p; % Synchronous speed
Zs=Rs+Xs*j; Zm=0+Xm/2*j;
for i=1:npts % Loop to calculate torque points
    ZU=Rr/2/s(i)+Xr/2*j; ZL=Rr/2/(2-s(i))+Xr/2*j;
    Zf=Zm*ZU/(Zm+ZU); Zb=Zm*ZL/(Zm+ZL);
    Is=V1/(Zs+Zf+Zb); I1(i)=abs(Is); nm(i)=(1-s(i))*ns;
    Irf=abs(Is*Zm/(Zm+ZU)); Irb=abs(Is*Zm/(Zm+ZL));
    PF(i)=cos(angle(Is)); Pin=V1*I1(i)*PF(i);
    Td(i)=0.5*Irf^2*Rr/s(i)/ws-0.5*Irb^2*Rr/(2-s(i))/ws;
    if Td(i)<0; Td(i)=0; end
    TPs=Td(i)*(1-s(i))*ws - Pfw*(nm(i)/ns)^n;
    if TPs<0; break; else;
        Ps(i)=TPs; eff(i)=100*Ps(i)/Pin;
    end
end
subplot(2,2,1), plot(nm,I1); grid; title('Input current');
xlabel('Speed, rpm'); ylabel('Current, A');

subplot(2,2,2); plot(nm,PF); grid; title('Input power factor');
xlabel('Speed, rpm'); ylabel('Power factor');

subplot(2,2,3), plot(nm,Td); grid; title('Developed torque');
xlabel('Speed, rpm'); ylabel('Torque, N-m');

subplot(2,2,4); plot(nm,eff); grid; title('Efficiency');
xlabel('Speed, rpm'); ylabel('Efficiency, %');
%*******************************************************************
```

PROGRAM OUTPUT

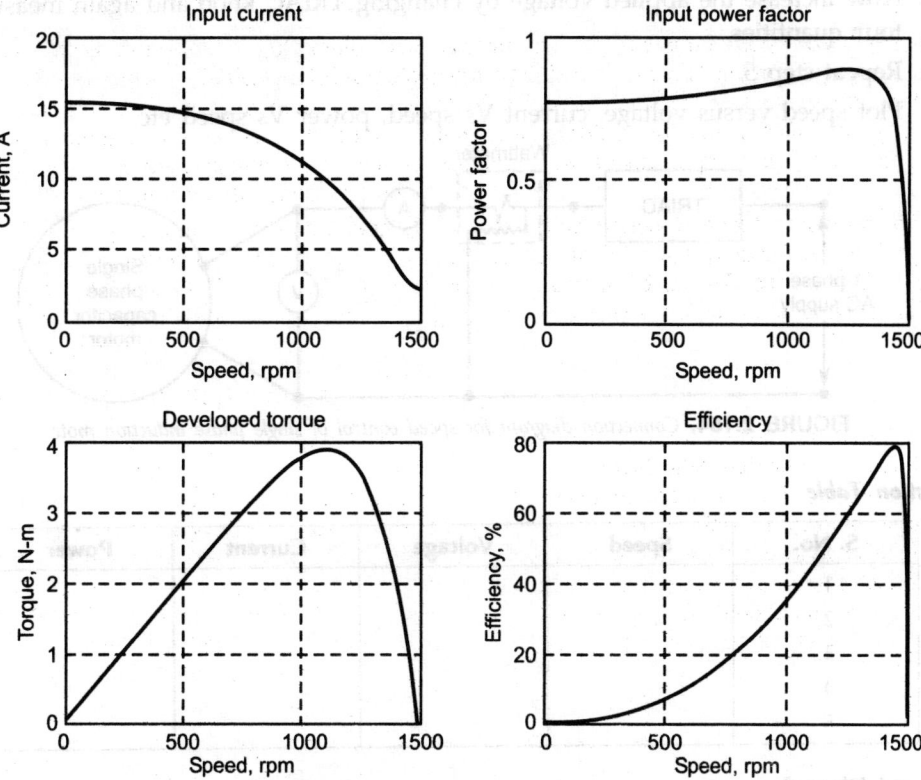

FIGURE 2.163. *Performance characteristics of single phase induction motor*

EXPERIMENT NO. 20

2.11.5 Speed Control of a Single Phase Motor using TRIAC

Need: There are many situations where it is required to control the speed of single phase induction motor. For example consider a single phase motor operated fan. Fan speed must be controlled depending on the temperature and humidity. Similarly in tunnels and in mines variable speed is required.

Apparatus

    Ammeter: 0 – 5A;

    Voltmeter: – 0 – 250 V.

    Wattmeter: – 0 –1400 W.

    Techometer

Procedure

1. Make the connections as shown in Figure 2.164.
2. Keep the TRIAC in maximum position.
3. Apply rated voltage (= 230 volts) to the terminals.

4. Measure speed, voltage, current and power on single phase motor.
5. Now increase the applied voltage by changing TRIAC knob and again measure all the four quantities.
6. Repeat step 5.
7. Plot speed versus voltage, current Vs speed, power Vs speed etc.

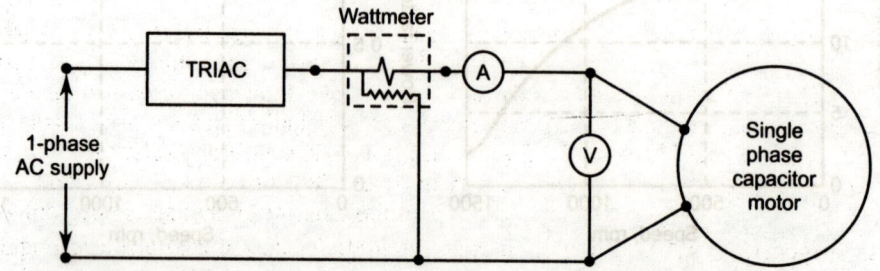

FIGURE 2.164. *Connection diagram for speed control of single phase induction motor*

Observation Table

S. No.	Speed	Voltage	Current	Power
1.				
2.				
3.				
4.				
5.				

Results and Discussion

Plot the performance characteristics speed versus voltage, current Vs speed, power Vs speed etc.

PROBLEM. *Write a Matlab program for simulating the speed-torque characteristics of single phase induction motor at different voltages and find the operating point.*

2.11.6 Matlab Program for Plotting Speed — Torque Characteristics of Single Phase Induction Motor for Different Voltages to Control its Speed

```
clear all;
for j1=1:16;
f=50;
p=16; % Phase voltage, frequency, poles
Rs=0.52;
Xs=0.79;
Xm=106.8;
Rr=6.12;
Xr=12.12;
Xc=7;
Pfw=1.5;
err= .2
n=2.8; % Total F&W losses at syn. speed, speed dependence
npts=100;
```

```
s=linspace( 0.0001,1,npts);
s=fliplr(s);
I1=zeros(1,npts);
Td=I1;
PF=I1;
Ps=I1;
eff=I1;
nm=I1;
ws=2/p*2*pi*f;
ns=120*f/p; % Synchronous speed
Zs=Rs+Xs*j;
Zm=0+Xm/2*j;
V1=70+j1*10;
for i=1:npts % Loop to calculate torque points
ZU=Rr/2/s(i)+Xr/2*j+(Xc)*(-j);
ZL=Rr/2/(2-s(i))+Xr/2*j;
Zf=Zm*ZU/(Zm+ZU);
Zb=Zm*ZL/(Zm+ZL);
Is=V1/(Zs+Zf+Zb);
I1(i)=abs(Is);
nm(i)=(1-s(i))*ns;
Irf=abs(Is*Zm/(Zm+ZU));
Irb=abs(Is*Zm/(Zm+ZL));
PF(i)=cos(angle(Is));
Pin=V1*I1(i)*PF(i);
Td(i)=0.5*Irf^2*Rr/s(i)/ws-0.5*Irb^2*Rr/(2-s(i))/ws;
Tl(i)=0.5+4.4e-4*nm(i).^2;

if Td(i)<0;
   Td(i)=0; end
TPs=Td(i)*(1-s(i))*ws - Pfw*(nm(i)/ns)^n;
if TPs<0;
   break;
else;
Ps(i)=TPs;
eff(i)=100*Ps(i)/Pin; end
if abs(Td(i)-Tl(i))<= err;
   n1=i
end
end
plot(nm,Tl);
hold on
title('Torque- Speed Curves for different voltages');
xlabel('Speed, rpm');
ylabel('Torque, N-m');
plot(nm,Td);
x1(j1,:)=[nm(n1); V1]'
end
```

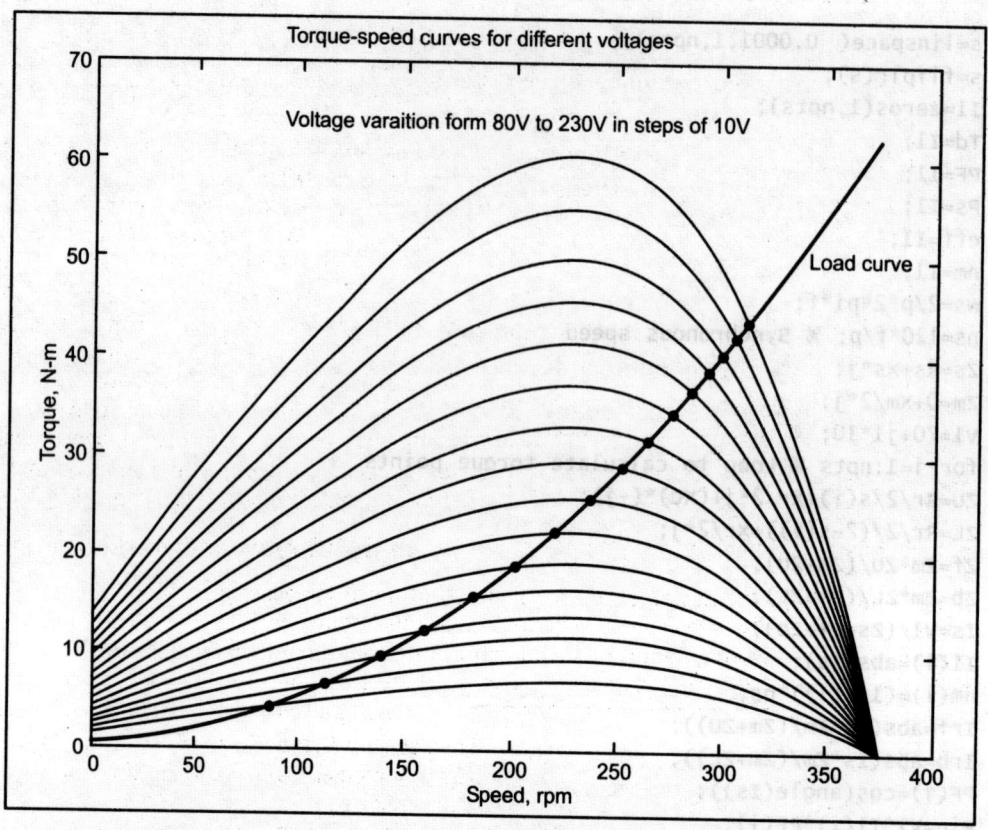

FIGURE 2.165. *Speed control of single phase induction motor for fan load*

2.11.7   Review Exercises

I. Multiple Choice Questions :

1. A single winding single phase motor is running in a particular direction
   (a) Both the rotating fields (rotating in opposite in directions) have the same strength.
   (b) Forward rotating field is slightly stronger than the backward field.
   (c) Forward rotating field is much stronger than the backward field.
   (d) Backward field is much slightly stronger than the forward field.

2. A shaded pole motor runs in
   (a) the direction from the shaded to unshaded part of the poles.
   (b) the direction from the unshaded to shaded part of the poles.
   (c) any direction depending on the polarity of the applied voltage.
   (d)   none of the above.

3. Compared to a capacitor start motor a two value capacitor motor has
   (a) nearly the same starting torque but better running power factor.
   (b) higher starting torque and higher running power factor
   (c) higher starting torque but lower running power factor
   (d) lower starting torque and lower running power factor.

4. A single phase reluctance motor
   (a) has salient pole rotor structure and run at synchronous speed
   (b) has salient pole rotor structure and run at sub - synchronous speed
   (c) has non-salient pole rotor structure and run at synchronous speed
   (d) has non-salient pole rotor structure and run at sub-synchronous speed

5. Starting torque in single phase hysteresis motor is caused by
   (a) Hysteresis   (b) Eddy current
   (c) Hysteresis and eddy current   (d) It has no starting torque.

6. A single phase reluctance motor
   (a) has zero starting torque as it is a synchronous motor
   (b) starts as induction motor but runs as synchronous motor
   (c) starts as hysteresis motor runs as synchronous motor
   (d) starts as induction motor but runs as induction motor.

7. A single winding single phase motor has
   (a) low starting torque   (b) zero starting torque
   (c) high starting torque   (d) starting torque equal to full load torque.

8. In a split phase motor, the running winding should have
   (a) high resistance and low inductance (b) low resistance and high inductance
   (c) high resistance and high inductance (d) low resistance and low inductance.

9. If the capacitor of a single phase motor is short circuited
   (a) motor will not start
   (b) the motor will burn
   (c) the motor will run in the reverse direction
   (d) the motor will run in the same direction at reduced speed.

10. In capacitor start and run motor the function of the running capacitor in series with the auxiliary winding is to
    (a) improve power factor   (b) increase overload capacity
    (c) reduce fluctuations in torque   (d) to improve torque.

11. Which of the following motor will give relative high starting torque
    (a) capacitor start motor   (b) capacitor run motor
    (c) split phase motor   (d) shaded pole motor.

12. Which type of capacitor is preferred for capacitor start and induction run motor
    (a) electrolyte capacitor   (b) oil capacitor
    (c) ceramic capacitor   (d) air capacitor.

13. For production of a rotating magnetic field
    (a) a single phase supply is to be connected across a single phase winding
    (b) a two-phase supply should be connected across a two phase winding
    (c) a dc supply is connected across a single phase winding
    (d) none of the above.

14. The direction of rotation of split phase induction motor can be reversed by
    (a) reversing the supply terminals
    (b) reversing the connection of auxiliary winding across the supply terminals
    (c) reversing the connection of auxiliary winding or main winding terminals
    (d) reversing the connection of only main winding across the supply terminals.

15. In a single phase repulsion motor, torque is developed on the rotor when the brush axis is fixed
    (a) at 90 degree electrical with the stator field axis
    (b) in alignment with the stator field axis
    (c) at an acute angle with the stator field axis
    (d) at 90 deg. mechanical with the stator field axis.

16. If the centrifugal switch of a resistance split phase induction motor fails to close when the motor is de-energised, then
    (a) no starting torque will be developed when supply is given
    (b) a dangerously high current will flow through motor
    (c) starting torque is not sufficient to start the motor
    (d) motor will develop high starting torque.

17. A dc series motor when connected across an ac supply will
    (a) develop torque in the same direction
    (b) not develop any torque
    (c) draw dangerously high current
    (d) develop pulsating torque.

18. To enable a dc series motor work satisfactorily with an ac supply, the following modifications should be done
    (a) the yoke and the pole should be laminated.
    (b) only the pole should be made of laminated sheets
    (c) the air gap between the stator and the rotor be reduced
    (d) compensating poles should be introduced.

II. True or False Statements :

1. The rotor of a split phase type single phase induction motor is provided with two windings.
2. The currents in main winding and auxiliary winding of a single phase induction motor are in phase with each other.
3. In shaded pole motor the starting torque is low.
4. The direction of shaded pole motor can not be reversed.
5. No winding is provided on the rotor of a hystersis motor.

Answers

I. Multiple Choice Questions :

1. (c)	2. (b)	3. (a)	4. (a)	5. (b)
6. (b)	7. (b)	8. (b)	9. (a)	10. (a)
11. (a)	12. (a)	13. (b)	14. (b)	15. (c)
16. (a)	17. (a)	18. (a)		

II. True or False Statements :

1. F	2. F	3. T	4. T	5. T

SAMPLE LAB QUIZ

SAMPLE LAB QUIZ-1

Q.1. Draw the circuit diagram of slip test of synchronous machine.   [2]

Q.2. Define: [1 × 3]

    (*i*) Sub-transient Reactance   (*ii*) Transient Reactance

    (*iii*) Steady State Reactance

Q.3. Name the relay which is gas actuated.   [1]

Q.4. Name the device, in which gas actuated relay is used.   [1]

Q.5. Synchronous reactance = ... reactance + ... Reactance.   [3]

Q.6. Ratio of X_d/X_q for cylindrical alternator is .... X_d/X_q of salient pole machine.   [1]

Q.7. Define: [1 × 3]

    (*i*) Positive Sequence Reactance

    (*ii*) Negative Sequence reactance

    (*iii*) Zero Sequence reactance

Q.8. Tick the correct answer: [1]

The armature voltage of a three phase alternator supplying power consists of ....

    (*i*) no sequence component   (*ii*) All the three sequence components

    (*iii*) only zero sequence component   (*iv*) only +ve sequence component

    (*v*) Only –ve sequence component

Q.9. Magnetic field set up in the armature of an alternator due to positive sequence currents rotates at .... speed in the direction .... to the direction of rotor.   [2]

Q.10. Zero sequence currents in the alternator produce a .... magnetic field in the stator. (Rotating / Pulsating).   [1]

Q.11. Draw the circuit diagram for an over voltage relay to draw its characteristic between time and voltage.   [2]

Q.1. Tick the most appropriate answer (Overwriting, cutting or erasing of any answer will be treated as wrong answer). [1 × 5]

    (*i*) The open circuit test of a transformer gives

        (*a*) Hysteresis loss   (*b*) Eddy current loss

        (*c*) sum of hysteresis and eddy current loss   (*d*) Copper loss

    (*ii*) Which of the following part is likely to suffer maximum damage due to excessive temperature rise

        (*a*) Core laminations   (*b*) Copper winding

        (*c*) Dielectric Strength of oil   (*d*) Winding insulation

    (*iii*) A transformer transforms

        (*a*) Energy   (*b*) Frequency

        (*c*) Power   (*d*) Voltage

        (*e*) all of the above.

    (*iv*) Which transformer will have smallest core

        (*a*) 1 kVA, 50 Hz   (*b*) 1 kVA, 100 Hz

        (*c*) 1 kVA, 200 Hz   (*d*) 1 kVA, 400 Hz.

    (*v*) The crawling in induction motor is caused by

        (*a*) Improper design of machine   (*b*) low voltage supply

        (*c*) High loads   (*d*) Harmonics developed in the motor.

Q.2. Answers in brief:

    1. Mention two advantages of delta-delta connection of transformers. [2]

    2. Mention the efficiency of transformer at same load at unity power factor and 0.8 lagging pf as calculated in the laboratory. Justify your answer. [2]

    3. What is the relationship between the mechanical degrees and electrical degrees? [1]

    4. In speed control of DC shunt motor, mention the range of ammeter connected in field circuit. [1]

    5. Name two protections, which are provided in a three-point starter of dc shunt motor. [2]

    6. What are the drawbacks of three-point starter, which are overcome, in four-point starter? [2]

    7. Draw the circuit diagram to plot open circuit characteristics of DC separately excited motor in the laboratory. [2]

    8. Draw open circuit and short circuit characteristics of a synchronous machine to calculate the voltage regulation. [2]

    9. Draw circuit diagram to perform the experiment of parallel operation of two single-phase transformers. [2]

    10. Mention the type of dc motor used for traction purposes. Draw external characteristics of traction motor. [2]

    11. How the induction motor is loaded in the laboratory. [2]

Additional Test Questions

    1. What happens if silicon content increases in silicon steel?

    2. What is the thickness of laminations used in Transformer?

3. What are insulating material for the windings if maximum temperature is less than or equal to 115°C?

4. Name the material for the bushing insulators for transformer?

5. What is dielectric strength of oil?

6. Draw the B-H characteristics for soft and hard magnetic materials?

7. What is type of material used for electromagnet?

8. What is type of material used for permanent magnet?

9. How the size of machine will change, if specific electric loading is increased?

10. Define the specific electric loading?

11. Mention the relationship between specific magnetic loading and pole pitch?

Solution:

1. Electrical Resistance Increases and hence, reduces the eddy currents and eddy current losses. If silicon content is greater than 3–4 % then it becomes brittle.

2. Varying from 0.3–0.5 mm.

3. Enameled wire on a base of polyvinyl formal, cotton, paper laminated with formaldehyde bonding.

4. Porcelain.

6. Soft magnetic material has B-H loop area lesser than hard magnetic material.

7. Soft magnetic material.

8. Hard magnetic material.

9. size of machine will reduce.

10. Specific electric loading $q = Ia\ Za\ /\ \pi D$.

11. Specific magnetic loading $B = \phi/\ (Y * L)$.

APPENDICES

APPENDIX-A

LAB RECORD FORMAT FOR EXPERIMENTS

Experiment No.

Date of Experiment   Date of Submission

  1.1 Objective

  1.2 Need of this experiment

  1.3 Instruments and devices required with specifications

  1.4 Procedural Steps

  1.5 Observation Table

  1.6 Calculations/Investigations

  1.7 Results/Conclusions

      Plot the results obtained and infer the results.

  1.8 Precautions (if any)

APPENDIX-B

FORMAT OF LAB RECORD EVALUATION BY LAB INSTRUCTOR

ELECTRICAL MACHINES LAB

Experiment No. .... Date of Experiment .... Date of Submission ....

Objective of Experiment: ....
....

Name of student .... S.No. ....

Class .... Group ....

Remarks:

Consultation needed	Report Incomplete
Redraw Diagram	Late Submission
Redraw Graphs	Re-compute Results
Repeat the experiment	

Marks/Grade ....

Signature of the Instructor

APPENDIX-C

UNITS AND THEIR CONVERSION FACTORS

Quantity	Units	Equivalent
Resistivity	1 ohm – meter (Ω-m)	10^2 Ohm-cm (Ω-cm)
		39.37 ohm - inches (Ω-in)
Magnetic Flux (Φ)	1 weber (wb)	10^8 Maxwells or lines
		10^5 kilo lines
Magnetic flux density (B)	1 tesla (T)	1 wb/m^2
		10^4 gauss
Magento Motive force (mmf)	1 ampere-turn (AT)	1.257 gilberts
Magnetic filed intensity (H)	1 ampere/meter(A/m)	2.54×10^{-2} A/in
		1.257×10^{-2} oersted
Angular velocity (ω)	1 radian /sec	$30/\pi$ rpm
		9.549 rpm
Linear velocity (v)	1 meter/sec (m/s)	3.6 Km/hr
Length (L)	1 meter (m)	39.37 in
	1 nautical mile	6080 ft
	1 yard	3 ft
	1 mile	1760 yard
Area (a)	1 square meter (m^2)	10^4 cm^2
		1550 in^2
		10.764 ft^2
	1 acre	4840 sq yards
		0.405 hactares
Volume (V)	1cubic meter (m^3)	10^6 cm^3
		35.315 ft^3
Mass (m)	1 kilogram (kg)	2.205 lb
		35.27 oz
Mass density	1 kilogram/meter^3 (kg/m^3)	6.243×10^{-2} lb/ft^3
		3.613×10^{-5} lb/in^3
Capacity	1 gallon (gal)	0.8327 UK gal
		3.7853 litres
Force (f)	1 newton (N)	1 m-Kg/s^2
		0.2248 pound (lb$_f$)

Torque (T)	1 newton –meter(N-m)	1.02×10^4 g-cm
		$0.738 \times$ lb-ft
		8.85 lb-in
Energy (E)	1 joule (J)	1 W-s
Power (P)	1watt (W)	1 J/s 1/746 hp
Current Density (δ)	1 ampere/meter^2 (A/m^2)	10^{-4} A/cm^2
Temperature (T)	1° Celcius (Centigrade)	(9/5°C +32) °F
		(°C +273) °K

APPENDIX-D

COLOR CODES FOR RESISTORS

TABLE 1. Color Codes for Resistor

Color	I-Band	II-Band	III-Band	IV – Band Tolerance, %
Black	0	0	1	
Brown	1	1	10	
Red	2	2	100	
Orange	3	3	1k	
Yellow	4	4	10k	
Green	5	5	100k	
Blue	6	6	1M	
Voilet	7	7	10M	
Gray	8	8	100M	
White	9	9	1G	
Gold			0.1	5%
Silver			0.001	10%
No Band				20%

Blank in Table means the situation does not exist.

TABLE 2. Comparison of Dielectric Constants of Capacitors

Dielectric	Dielectric constant, K
Air or Vacuum	1.0
Paper	2.0-6.0
Plastic	2.1-6.0
Mineral oil	2.2-2.3
Silicone oil	2.7-2.8
Quartz	3.8-.4.4
Glass	4.8-8.0
Porcelain	5.1-5.9
Mice	5.4-8.7
Aluminium oxide	8.4
Tantalum pentoxide	26
Ceramic	12-400,000

TABLE 3. Different Machines and their Applications

Electrical Machines	Applications
1. Transformers	
Distribution transformer	At distribution stations, large scale industries, near colonies to step down the voltage.
Power transformer	At generating stations to step up the voltage
Instrument transformer	To extend the range of voltmeter, ammeter and wattmeter.
Welding transformer	For welding purpose
2. DC machines	
i. Seperately excited	DC power supply purpose
ii. Shunt Generators	a. Battery Charging
	b. Pilot exciter for synchronous generator
iii. Series Generators	a. Booster
	b. Constant current source in welding
iv. Compound generators	DC supply at constant voltage
v. Shunt motor	Wood working machine, Battery operated fans, solar fans
vi. Series motor	Electric locomotives, starter motor in automobiles
vii. compound motors	Rolling mills and Printing machines
3. AC machines	
i. Induction motors	
a. 3-phase squirrel cage	Lathes, drilling machines, agricultural and industrial pumps, loom motors, compressors and inductrial drive.
b. 3-phase slip ring motors	Lift, Crane, Conveyor.
c. 1-phase induction motor	
i. capacitor start and run	Grinder, domestic pumps, fans, washing and refrigerator motors.
ii. Shaded pole	Hair dryer, sewing machines
d. 3-phase induction generators	Windmills
4. Synchronoous machines	
i. Generator	Power stations, stand by alternators, battery charger in automobiles
ii. Motor	Constant speed applications, rotating condensers for power factor improvements.

APPENDIX-E

BASICS OF MATLAB

1. WHAT IS MATLAB?

MATLAB is a high-performance language for technical computing. It integrates computation, visualization, and programming in an easy-to-use environment where problems and solutions are expressed in familiar mathematical notation. Typical uses include :

- Math and computation
- Algorithm development
- Modeling, simulation, and prototyping
- Data analysis, exploration, and visualization
- Scientific and engineering graphics
- Application development, including graphical user interface building

MATLAB is an interactive system whose basic data element is an array that does not require dimensioning. This allows to solve many technical-computing problems, especially those with matrix and vector formulations, in a fraction of the time it would take to write a program in a scalar noninteractive language such as C or Fortran. The name MATLAB stands for *matrix laboratory*. In industry, MATLAB is the tool of choice for high-productivity research, development, and analysis. MATLAB features a family of application-specific solutions called *toolboxes*. Very important to most users of MATLAB, toolboxes allow to *learn* and *apply* specialized technology. Toolboxes are comprehensive collections of MATLAB functions (M-files) that extend the MATLAB environment to solve particular classes of problems. Areas in which toolboxes are available include signal processing, control systems, neural networks, fuzzy logic, wavelets, simulation, and many others.

2. LEARNING MATLAB

Learning MATLAB is just like learning how to drive a car. It is possible to learn all the rules from the manual but to become a good driver we have to get out on the road and drive. **The easiest and best way to learn MATLAB is to use MATLAB.**

3. THE MATLAB SYSTEM

The MATLAB system consists of five main parts:

3.1 Development Environment

This is the set of tools and facilities that help you use MATLAB functions and files. Many of these tools are graphical user interfaces. It includes the MATLAB desktop and Command Window, a command history, and browsers for viewing help, the workspace, files, and the search path.

3.2 The MATLAB Mathematical Function Library

This is a vast collection of computational algorithms ranging from elementary functions like sum, sine, cosine, and complex arithmetic, to more sophisticated functions like matrix inverse, matrix Eigen values, Bessel functions, and fast Fourier transforms.

3.3 The MATLAB Language

This is a high-level matrix/array language with control flow statements, functions, data structures, input/output, and object-oriented programming features. It allows both "programming in the small" to rapidly create quick and dirty throw-away programs, and "programming in the large" to create complete large and complex application programs.

3.4 Handle Graphics

The MATLAB includes high-level commands for two-dimensional and three-dimensional data visualization, image processing, animation, and presentation of graphics. It also includes low-level commands that allow you to fully customize the appearance of graphics as well as to build complete graphical user interfaces.

3.5 The MATLAB Application Program Interface (API)

This is a library that allows us to write C and Fortran programs that interact with MATLAB. It include facilities for calling routines from MATLAB (dynamic linking), calling MATLAB as a computational engine, and for reading and writing MAT-files.

4. STARTING AND QUITTING MATLAB

On a MS-Windows platform, to start MATLAB, double-click the MATLAB shortcut icon on Windows desktop. On a UNIX platform, to start MATLAB, type matlab at the operating system prompt.

To end your MATLAB session, select **Exit MATLAB** from the **File** menu in the desktop, or type quit in the Command Window. To execute specified functions each time MATLAB quits, such as saving the workspace, we can create and run a .m script.

FIGURE A1. *Desktop with Matlab icon*

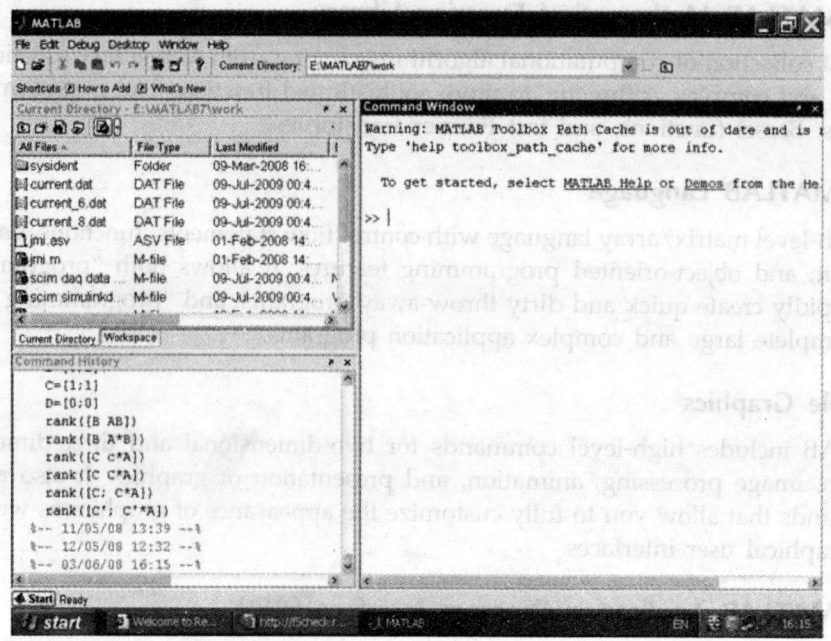

FIGURE A2. *Matlab Window*

5. MATLAB DESKTOP

When we start MATLAB, the MATLAB desktop appears, containing tools (graphical user interfaces) for managing files, variables, and applications associated with MATLAB. The first time MATLAB starts, the desktop appears as shown in fig. A3, although Launch Pad may contain different entries. We can change the desktop looks by opening, closing, moving, and resizing the tools in it. We can also move tools outside of the desktop or return them back inside the desktop (docking). All the desktop tools provide common features such as context menus and keyboard shortcuts. We can specify certain characteristics for the desktop tools by selecting **Preferences** from the **File** menu. For example, we can specify the font characteristics for Command Window text. For more information, click the **Help** button in the **Preferences** dialog box.

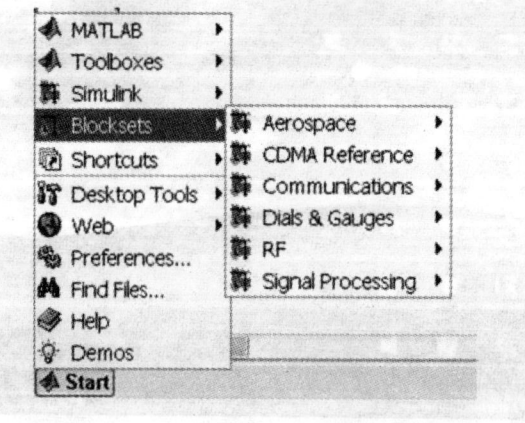

6. DESKTOP TOOLS

This section provides an introduction to MATLAB's desktop tools. We can also use MATLAB functions to perform most of the features found in the desktop tools.

6.1 Command Window

Use the **Command Window** to enter variables and run functions and M-files.

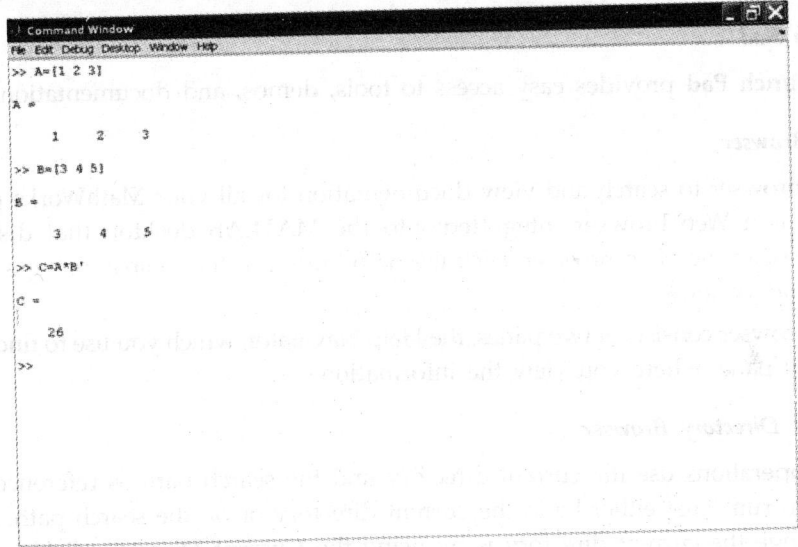

6.2 Command History

Lines enter in the Command Window are logged in the **Command History** window. In the Command History, one can view previously used functions, and copy and execute selected lines.

6.2.1 Running External Programs

We can run external programs from the MATLAB Command Window. The exclamation point character ! is a shell escape and indicates that the rest of the input line is a command to the operating system. This is useful for invoking utilities or running other programs without quitting MATLAB. On Linux, for example, !emacs magik.m invokes an editor called emacs for a file named magik.m. When we quit the external program, the operating system returns control to MATLAB.

6.2.2 Launch Pad

MATLAB's **Launch Pad** provides easy access to tools, demos, and documentation.

6.2.3 Help Browser

Use the Help browser to search and view documentation for all your MathWorks products. The Help browser is a Web browser integrated into the MATLAB desktop that displays HTML documents. To open the Help browser, click the help button in the toolbar, or type help browser in the Command Window.

The Help browser consists of two panes, the Help Navigator, which you use to find information, and the display pane, where you view the information.

6.2.4 Current Directory Browser

MATLAB file operations use the current directory and the search path as reference points. Any file you want to run must either be in the current directory or on the search path. A quick way to view or change the current directory is by using the **Current Directory** field in the desktop toolbar as shown below.

To search for, view, open, and make changes to MATLAB-related directories and files, use the MATLAB Current Directory browser. Alternatively, you can use the functions dir, cd, and delete.

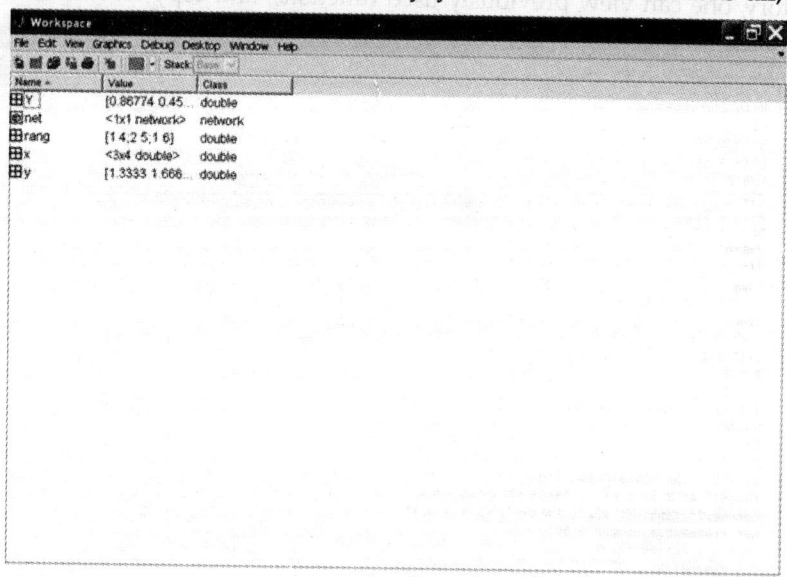

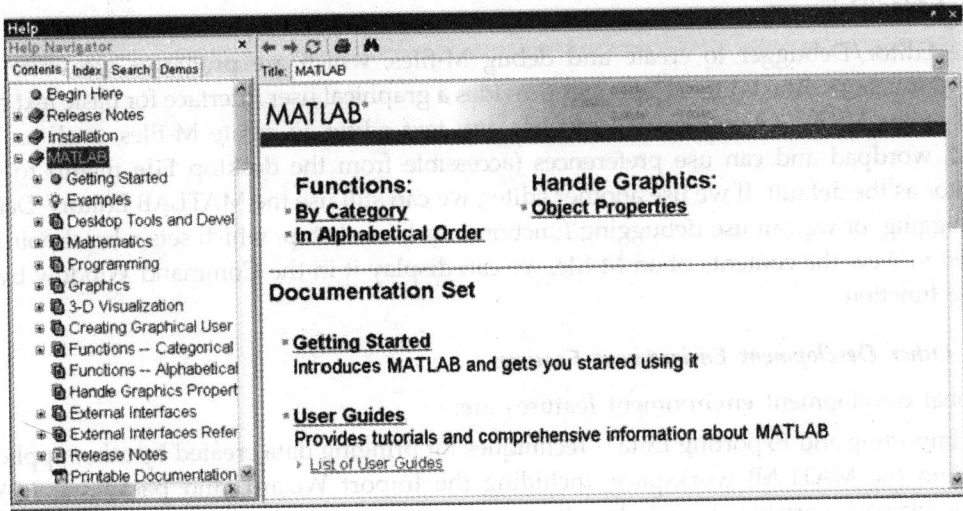

6.2.5 Workspace Browser

The MATLAB workspace consists of the set of variables (named arrays) built up during a MATLAB session and stored in memory. We add variables to the workspace by using functions, running M-files, and loading saved workspaces. To view the workspace and information about each variable, use the Workspace browser, or use the functions who and whos. To delete variables from the workspace, select the variable and select **Delete** from the **Edit** menu. Alternatively, use the clear function. The workspace is not maintained after ending of the MATLAB session. To save the workspace to a file that can be read during a later MATLAB session, select **Save Workspace As** from the **File** menu, or use the save function. This saves the workspace to a binary file called a MAT-file, which has a .mat extension. There are options for saving to different formats. To read in a MAT-file, select **Import Data** from the **File** menu, or use the load function.

6.2.6 Array Editor

Double-click on a variable in the Workspace browser to see it in the Array Editor. Use the Array Editor to view and edit a visual representation of one- or two-dimensional numeric arrays, strings, and cell arrays of strings that are in the workspace. Change values of array elements. Change the display format.

6.2.7   Editor/Debugger

Use the Editor/Debugger to create and debug M-files, which are programs we write to run MATLAB functions. The Editor/Debugger provides a graphical user interface for basic text editing, as well as for M-file debugging. We can use any text editor to create M-files, such as Emacs, notepad, wordpad and can use preferences (accessible from the desktop **File** menu) to specify that editor as the default. If we use another editor, we can still use the MATLAB Editor/ Debugger for debugging, or we can use debugging functions, such as dbstop, which sets a breakpoint. If we just need to view the contents of an M-file, we can display it in the Command Window by using the type function.

6.2.8   Other Development Environment Features

Additional development environment features are:

- Importing and Exporting Data – Techniques for bringing data created by other applications into the MATLAB workspace, including the Import Wizard, and packaging MATLAB workspace variables for use by other applications.
- Improving M-File Performance – The Profiler is a tool that measures where an M-file is spending its time. Use it to help you make speed improvements.
- Interfacing with Source Control Systems – Access your source control system from within MATLAB, Simulink, and State flow.

7. ENTERING MATRICES

The best way to get started with MATLAB is to learn how to handle matrices. Start MATLAB and follow along with each example. Matrices can be entered into MATLAB in several different ways:

- Enter an explicit list of elements.
- Load matrices from external data files.
- Generate matrices using built-in functions.
- Create matrices with your own functions in M-files.

Start by entering matrix as a list of its elements. You have only to follow a few basic conventions:

- Separate the elements of a row with blanks or commas.
- Use a semicolon, ; , to indicate the end of each row.
- Surround the entire list of elements with square brackets, [ ]. To enter matrix, simply type in the Command Window

$$A = [16\ 3\ 2\ 13;\ 5\ 10\ 11\ 8;\ 9\ 6\ 7\ 12;\ 4\ 15\ 14\ 1]$$

Or

$$A = \begin{bmatrix} 16 & 3 & 2 & 13 \\ 5 & 10 & 11 & 8 \\ 9 & 6 & 7 & 12 \\ 4 & 15 & 14 & 1 \end{bmatrix}$$

MATLAB displays the matrix just entered as:

$$A = \begin{matrix} 16 & 3 & 2 & 13 \\ 5 & 10 & 11 & 8 \\ 9 & 6 & 7 & 12 \\ 4 & 15 & 14 & 1 \end{matrix}$$

Once the matrix entered, it is automatically remembered in the MATLAB workspace. We can refer to it simply as A. Now you have A in the workspace, take a look at what makes it so interesting. Why is it magic?

sum, transpose, and diag

sum, transpose, and diag

When magic square matrix elements have summed up in various ways, it gives same number. If you take the sum along any row or column, or along either of the two main diagonals, you will always get the same number. Let's verify that using MATLAB. The first statement to try is sum(A) MATLAB replies with

$$>> ans = 34 \quad 34 \quad 34 \quad 34$$

When you don't specify an output variable, MATLAB uses the variable ans, short for *answer*, to store the results of a calculation. You have computed a row vector containing the sums of the columns of A. Sure enough, each of the columns has the same sum, the *magic* sum, 34. How about the sum of rows? MATLAB has a preference for working with the columns of a matrix, so the easiest way to get the row sums is to transpose the matrix, compute the column sums of the transpose, and then transpose the result. The transpose operation is denoted by an apostrophe or single quote, '. It flips a matrix about its main diagonal and it turns a row vector into a column vector. So **A'** produces

$$ans = \begin{matrix} 16 & 5 & 9 & 4 \\ 3 & 10 & 6 & 15 \\ 2 & 11 & 7 & 14 \\ 13 & 8 & 12 & 1 \end{matrix}$$

and

sum(A')' produces a column vector containing the row sums

$$ans = \begin{array}{c} 34 \\ 34 \\ 34 \\ 34 \end{array}$$

The sum of the elements on the main diagonal is easily obtained with the help of the diag function, which picks off that diagonal.

diag(A) produces

$$ans = \begin{array}{c} 16 \\ 10 \\ 7 \\ 1 \end{array}$$

and **sum(diag(A))** produces

$$ans = 34$$

The other diagonal, the so-called *antidiagonal,* is not so important mathematically, so MATLAB does not have a ready-made function for it. But a function originally intended for use in graphics, fliplr, flips a matrix from left to right.

```
>> sum(diag(fliplr(A)))
```
$$ans = 34$$

8. SUBSCRIPTS

The element in row i and column j of A is denoted by $A(i, j)$. For example, $A(4, 2)$ is the element in the fourth row and second column. For our magic square, $A(4, 2)$ is 15. So it is possible to compute the sum of the elements in the fourth column of A by typing

$$A(1,4) + A(2, 4) + A(3, 4) + A(4, 4)$$

This produces $ans = 34$

but is not the most elegant way of summing a single column. It is also possible to refer to the elements of a matrix with a single subscript, $A(k)$. This is the usual way of referencing row and column vectors. But it can also apply to a fully two-dimensional matrix, in which case the array is regarded as one long column vector formed from the columns of the original matrix. So, for our magic square, $A(8)$ is another way of referring to the value 15 stored in $A(4, 2)$. If we try to use the value of an element outside of the matrix, it is an error.

$$t = A(4, 5)$$

Index exceeds matrix dimensions.

On the other hand, if you store a value in an element outside of the matrix, the size increases to accommodate the newcomer.

$$X = A;$$
$$X(4, 5) = 17$$

$$X = \begin{array}{ccccc} 16 & 3 & 2 & 13 & 0 \\ 5 & 10 & 11 & 8 & 0 \\ 9 & 6 & 7 & 12 & 0 \\ 4 & 15 & 14 & 1 & 17 \end{array}$$

9. THE COLON OPERATOR

The colon, :, is one of MATLAB's most important operators. It occurs in several different forms.

(*i*) The expression 1:10 is a row vector containing the integers from 1 to 10.

>>1:10

>>ans = 1   2   3   4   5   6   7   8   9   10

To obtain non unit spacing, specify an increment. For example, 100:-7:50 is

>> X = 100:-7:50

>> X = 100 93 86 79 72 65 58 51

and

>> Angle = 0:pi/4:pi

>> Angle = 0 0.7854   1.5708   2.3562   3.1416

(*ii*) Subscript expressions involving colons refer to portions of a matrix. $A(1 : k, j)$ is the first k elements of the jth column of A. So sum($A(1 : 4, 4)$) computes the sum of the fourth column. But there is a better way. The colon by itself refers to *all* the elements in a row or column of a matrix and the keyword end refers to the *last* row or column. So sum($A(:, end)$) computes the sum of the elements in the last column of A.

>> ans = 34

Why is the magic sum for a 4-by-4 square equal to 34? If the integers from 1 to 16 are sorted into four groups with equal sums, that sum must be sum(1:16)/4 which, of course, is

ans = 34

10. THE MATLAB FUNCTIONS

MATLAB actually has a built-in function that creates magic squares of almost any size. Not surprisingly, this function is named magic.

>> B = magic(4)

$$
>> B = \begin{matrix} 16 & 2 & 3 & 13 \\ 5 & 11 & 10 & 8 \\ 9 & 7 & 6 & 12 \\ 4 & 14 & 15 & 1 \end{matrix}
$$

Besides this there are various other built in functions to generate commonly used matrices such as eye, ones, zeros and rand, etc.

>> eye (4)

$$
>> ans = \begin{matrix} 1 & 0 & 0 & 0 \\ 0 & 1 & 0 & 0 \\ 0 & 0 & 1 & 0 \\ 0 & 0 & 0 & 1 \end{matrix}
$$

```
>> ones (4)

        1   1   1   1
        1   1   1   1
>> ans =
        1   1   1   1
        1   1   1   1

>> zeros (3)

        0   0   0
>> ans = 0   0   0
        0   0   0

>> rand (3)

        0.9501   0.4860   0.4565
>> ans = 0.2311   0.8913   0.0185
        0.6068   0.7621   0.8214

>> randn (3)

        -0.4326   0.2877   1.1892
>> ans = -1.6656  -1.1465  -0.0376
        0.1253    1.1909   0.3273

>> N = fix(10*rand(1,10))

>> N = 4   9   4   4   8   5   2   6   8   0
```

11. EXPRESSIONS

Like most other programming languages, MATLAB provides mathematical *expressions*, but unlike most programming languages, these expressions involve entire matrices. The building blocks of expressions are:

- Variables
- Numbers
- Operators, and
- Functions

11.1 Variables

MATLAB does not require any type declarations or dimension statements. When MATLAB encounters a new variable name, it automatically creates the variable and allocates the appropriate amount of storage. If the variable already exists, MATLAB changes its contents and, if necessary, allocates new storage. For example, num_students = 25 creates a 1-by-1 matrix named num_students and stores the value 25 in its single element. Variable names consist of a letter, followed by any number of letters, digits, or underscores. MATLAB uses only the first 31 characters of a variable name. MATLAB is case sensitive; it distinguishes between uppercase and lowercase letters. A and a are *not* the same variable. To view the matrix assigned to any variable, simply enter the variable name.

11.2 Numbers

MATLAB uses conventional decimal notation, with an optional decimal point and leading plus or minus sign, for numbers. *Scientific notation* uses the letter e to specify a power-of-ten scale factor. *Imaginary numbers* use either i or j as a suffix. Some examples of legal numbers are any positive or negative real or integer number, for example –

3	–99	0.0001
9.6397238	1.60210e-20	6.02252e23
1i	–3.14159*j	3e5*i

All numbers are stored internally using the *long* format specified by the IEEE floating-point standard. Floating-point numbers have a finite *precision* of roughly 16 significant decimal digits and a finite *range* of roughly 10^{-308} to 10^{+308}.

11.3 Operators

Expressions use familiar arithmetic operators like +, –, *, /, \ , ^, ./, .\ etc. and precedence rules.

11.4 Functions

MATLAB provides a large number of standard elementary mathematical functions, including abs, sqrt, exp, and sin. Taking the square root or logarithm of a negative number is not an error; the appropriate complex result is produced automatically. MATLAB also provides many more advanced mathematical functions, including Bessel and gamma functions. Most of these functions accept complex arguments. For a list of the elementary functions type

```
help elfun
```

For a list of more advanced mathematical and matrix functions, type

```
helps specfun
```

Some of the functions, like sqrt and sin, are *built-in*. They are part of the MATLAB core so they are very efficient, but the computational details are not readily accessible. Other functions, like gamma and sinh, are implemented in M-files. You can see the code and even modify it if you want. Several special functions provide values of useful constants. Infinity is generated by dividing a nonzero value by zero, or by evaluating well defined mathematical expressions that *overflow, i.e.,* exceed realmax. Not a number is generated by trying to evaluate expressions like 0/0 or ∞-∞ or ∞/∞, etc. that do not have well defined mathematical values. The function names are not reserved. It is possible to overwrite any of them with a new variable, such as eps = 1.e-6 and then use that value in subsequent calculations.

pi	3.14159265…
i	Imaginary unit, $\sqrt{-1}$
j	Same as *i*
eps	Floating-point relative precision, 2^{-52}
realmin	Smallest floating-point number, 2^{-1022}
realmax	Largest floating-point number, $(2-\varepsilon)^{1023}$
Inf	Infinity (∞)
NaN	Not-a-number

12. THE LOAD COMMAND

The load command reads binary files containing matrices generated by earlier MATLAB sessions, or reads text files containing numeric data. The text file should be organized as a rectangular table of numbers, separated by blanks, with one row per line, and an equal number of elements in each row. Store the file under the name **filename.dat**. Then the command **load filename.dat** reads the file and creates a variable, x.

M-Files: You can create your own matrices using *M-files*, which are text files containing MATLAB code. Use the MATLAB Editor or another text editor to create a file containing the same statements which would type at the MATLAB command line. Save the file under a name that ends in .m. Store the file under the name **filename.m**. Then the statement **filename** reads the file and creates a variable, contained by the file.

Concatenation: *Concatenation* is the process of joining small matrices to make bigger ones. In fact, you made your first matrix by concatenating its individual elements. The pair of square brackets, [], is the concatenation operator. For an example, start with the 4-by-4 magic square, A, and form $B = [A \ A + 32; A + 48 \ A + 16]$

Deleting Rows and Columns: You can delete rows and columns from a matrix using just a pair of square brackets. Start with

>> X = A;

Then, to delete the second column of X, use

>> X(:,2) = []

If you delete a single element from a matrix, the result isn't a matrix anymore. So, expressions like $X(1,2) = [\ ]$ result in an error.

Adding a matrix to its transpose produces a *symmetric* matrix.

>> A + A'

The determinant of this particular matrix happens to be zero, indicating that the matrix is *singular*.

>> d = det(A)	
>> X = inv(A);	% Inverse of Matrix A.
>> RCOND = 1.175530e-017.	
>> [e, v] = eig(A);	% Eigen values and vector of Matrix A.
>> poly(A);	%the coefficients in the characteristic polynomial
>> mu = mean(D);	% Average Value of each column of matrix D
>> sigma = std(D);	% Standard deviation of each column of matrix D
>> B = A – 8.5 A;	% scalar is subtracted from each element of matrix A.

Controlling Command Window Input and Output: This section describes how to:

- Control the appearance of the output values
- Suppress output from MATLAB commands
- Enter long commands at the command line
- Edit the command line

13. THE FORMAT COMMAND

The format command controls the numeric format of the values displayed by MATLAB. The command affects only how numbers are displayed, not how MATLAB computes or saves them.

```
>> x = [4/3 1.2345e-6]
>> format short
>> ans =
        1.3333 0.0000
```

format short e
        1.3333e+000 1.2345e-006

format short g
        1.3333 1.2345e-006

format long
        1.33333333333333 0.00000123450000

format long e
        1.333333333333333e+000 1.234500000000000e-006

format long g
        1.33333333333333 1.2345e-006

format bank
        1.33 0.00

format rat
        4/3 1/810045

format hex
        3ff5555555555555 3eb4b6231abfd271

If you want more control over the output format, use the sprintf and fprintf functions.

14. SUPPRESSING OUTPUT

Type a statement on command window and press **Return** or **Enter**, MATLAB automatically displays the results on screen. However, if you end the line with a semicolon (;), MATLAB performs the computation but does not display any output. This is particularly useful when you generate large matrices. For example,

$$A = \text{magic}(100);$$

15. ENTERING LONG COMMAND LINES

If a statement does not fit on one line, use three periods, ..., followed by **Return** or **Enter** to indicate that the statement continues on the next line. For example,

$$s = 1 - 1/2 + 1/3 - 1/4 + 1/5 - 1/6 + 1/7 \ldots$$
$$- 1/8 + 1/9 - 1/10 + 1/11 - 1/12;$$

16. BASIC PLOTTING

MATLAB has extensive facilities for displaying vectors and matrices as graphs, as well as annotating and printing these graphs. This section describes a few of the most important graphics functions and provides examples of some typical applications.

16.1 Creating a Plot

The plot function has different forms, depending on the input arguments. If y is a vector, plot(y) produces a piecewise linear graph of the elements of y versus the index of the elements of y. If you specify two vectors as arguments, plot (x, y) produces a graph of y versus x. For example, these statements use the colon operator to create a vector of x values ranging from zero to 2π, compute the sine of these values, and plot the result.

```
>> x = 0:pi/100:2*pi;
>> y = sin(x);
>> plot(x, y)
```

Now label the axes and add a title. The characters \pi create the symbol π.

```
>> xlabel('x = 0:2\pi')
>> ylabel('Sine of x')
>> title('Plot of the Sine Function','FontSize',12)
```

16.2 Multiple Data Sets in One Graph

Multiple x-y pair arguments create multiple graphs with a single call to plot. MATLAB automatically cycles through a predefined (but user settable) list of colors to allow discrimination between each set of data. For example, these statements plot three related functions of x, each curve in a separate distinguishing color.

```
>> y2 = sin(x-.25);
>> y3 = sin(x-.5);
>> plot(x,y,x,y2,x,y3)
>> legend('sin(x)','sin(x-.25)','sin(x-.5)')
```

The legend command provides an easy way to identify the individual plots.

16.3 Plotting Lines and Markers

If only the markers are needed to plot of the graphics screen without connecting line, then the following command may be given.

```
>> plot(x,y,'ks- ')
```

ks – black squares at each data point connected with solid line.

The first character is colour,

Second character is marker type and

Third character for line type. These may be used in plot command in different combinations. The different colours, marker types and line types may be used are shown in Table 4.

TABLE 4. Different Colours, Marker Type and Line Types

Colours	Marker Types	Line Types
b blue	. point	- solid
g green	o circle	: dotted
r red	x x-mark	-. Dashdot
c cyan	+ plus	-- dashed
m magenta	* star	(none) no line
y yellow	s square	
k black	d diamond	
	v triangle (down)	
	^ triangle (up)	
	< triangle (left)	
	> triangle (right)	
	p pentagram	
	h hexagram	

```
plot(x,y,'s')     Plot the squares on the graph
plot(x,y,'s-')    Plot the squares connected with solid lines on the graph
plot(x,y,'rs-')   plot the squares connected with solid lines on the graph in red
                  colour.
```

The following example plots the data twice using a different number of points for the dotted line and marker plots.

```
x1 = 0:pi/100:2*pi;
x2 = 0:pi/10:2*pi;
plot(x1,sin(x1),'r:',x2,sin(x2),'r+')
```

Imaginary and Complex Data: When the arguments to plot are complex, the imaginary part is ignored *except* when plot is given a single complex argument. For this special case, the command is a shortcut for a plot of the real part versus the imaginary part. Therefore, plot(Z) where Z is a complex vector or matrix, is equivalent to plot(real(Z),imag(Z))

For example,

```
t = 0:pi/10:2*pi;
plot(exp(i*t),'-o')
axis equal
```

16.4 Adding Plots to an Existing Graph

The hold command enables you to add plots to an existing graph. When you type hold on MATLAB does not replace the existing graph when you issue another plotting command; it adds the new data to the current graph, rescaling the axes if necessary. For example, these statements first create a contour plot of the peaks function, then superimpose a pseudocolor plot of the same function.

```
[x,y,z] = peaks;
contour(x,y,z,20,'k')
hold on
pcolor(x,y,z)
shading interp
hold off
```

The hold on command causes the pcolor plot to be combined with the contour plot in one figure.

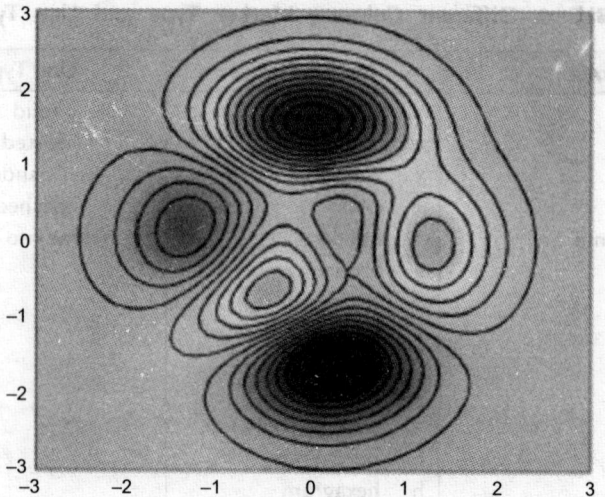

16.5   Multiple Plots in One Figure

The subplot command enables you to display multiple plots in the same window or print them on the same piece of paper. Typing

    subplot(m,n,p)

partitions the figure window into an m-by-n matrix of small subplots and selects the pth subplot for the current plot. The plots are numbered along first the top row of the figure window, then the second row, and so on. For example, these statements plot data in four different sub regions of the figure window as shown below.

```
t = 0:pi/10:2*pi;
[X,Y,Z] = cylinder(4*cos(t));
subplot(2,2,1);  mesh(X)
subplot(2,2,2);  mesh(Y)
subplot(2,2,3);  mesh(Z)
subplot(2,2,4);  mesh(X,Y,Z)
```

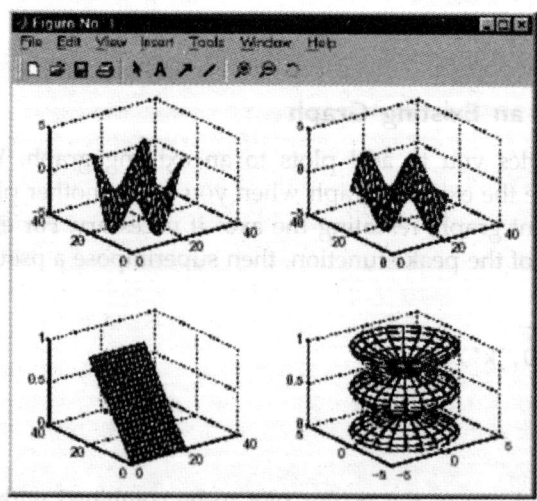

FIGURE A3. *Graphics screen*

16.6 Setting Grid Lines

The grid command toggles grid lines on and off. The statement

grid on   turns the grid lines on and

grid off   turns them back off again.

16.7 Axis Labels and Titles

The xlabel, ylabel, and zlabel commands add x-, y-, and z-axis labels. The title command adds a title at the top of the figure and the text function inserts text anywhere in the figure. A subset of TeX notation produces Greek letters. You can also set these options interactively.

```
t = -pi:pi/100:pi;
y = sin(t);
plot(t,y) axis([-pi pi -1 1])
xlabel('-\pi \leq {\itt} \leq \pi')
ylabel('sin(t)')
title('Graph of the sine function')
text(1,-1/3,'{\itNote the odd symmetry.}')
```

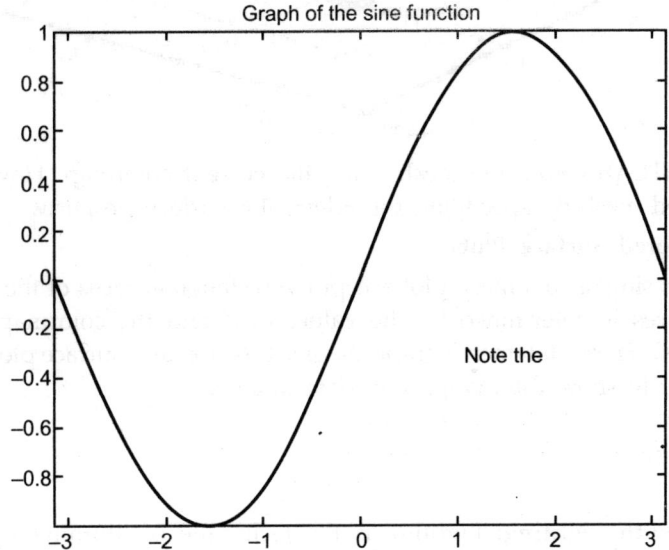

16.8 Saving a Figure

To save a figure, select **Save** from the **File** menu. To save it using a graphics format, such as TIFF, for use with other applications, select **Export** from the **File** menu. You can also save from the command line – use the saveas command, including any options to save the figure in a different format.

16.9 Mesh and Surface Plots

MATLAB defines a surface by the z-coordinates of points above a grid in the x-y plane, using straight lines to connect adjacent points. The **mesh** and **surf** plotting functions display surfaces in three dimensions. **mesh** produces wireframe surfaces that color only the lines connecting the defining points. **surf** displays both the connecting lines and the faces of the surface in color.

```
[X,Y] = meshgrid(-8:.5:8);
R = sqrt(X.^2 + Y.^2) + eps;
Z = sin(R)./R;
mesh(X,Y,Z,'EdgeColor','black')
```

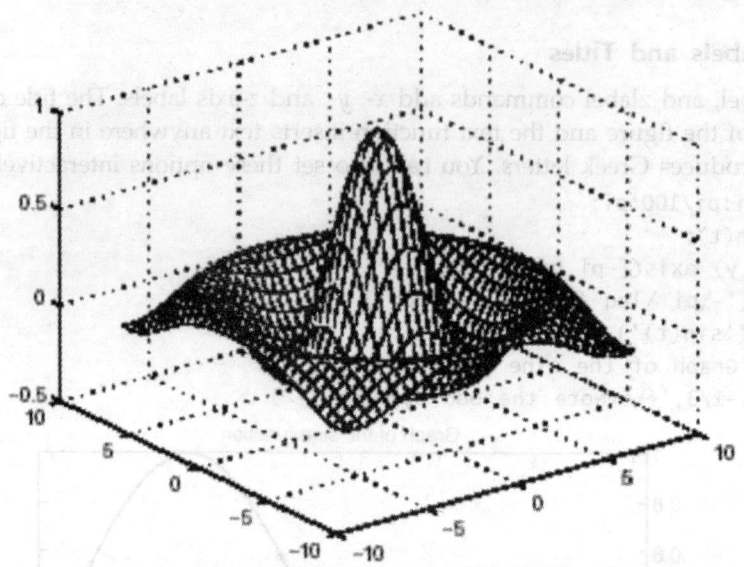

By default, MATLAB colors the mesh using the current colormap. However, this example uses a single-colored mesh by specifying the EdgeColor surface property.

Example – Colored Surface Plots

A surface plot is similar to a mesh plot except the rectangular faces of the surface are colored. The color of the faces is determined by the values of Z and the colormap (a colormap is an ordered list of colors). These statements graph the *sinc* function as a surface plot, select a colormap, and add a color bar to show the mapping of data to color.

```
surf(X,Y,Z)
colormap hsv
colorbar
```

Surface Plots with Lighting: Lighting is the technique of illuminating an object with a directional light source. In certain cases, this technique can make subtle differences in surface shape easier to see. Lighting can also be used to add realism to three-dimensional graphs. This example uses the same surface as the previous examples, but colors it red and removes the mesh lines. A light object is then added to the left of the"camera" (that is the location in space from where you are viewing the surface). After adding the light and setting the lighting method to phong, use the view command to change the view point so you are looking at the surface from a different point in space (an azimuth of -15 and an elevation of 65 degrees). Finally, zoom in on the surface using the toolbar zoom mode.

```
surf(X,Y,Z,'FaceColor','red','EdgeColor','none');
camlight left; lighting phong
view(-15,65)
```

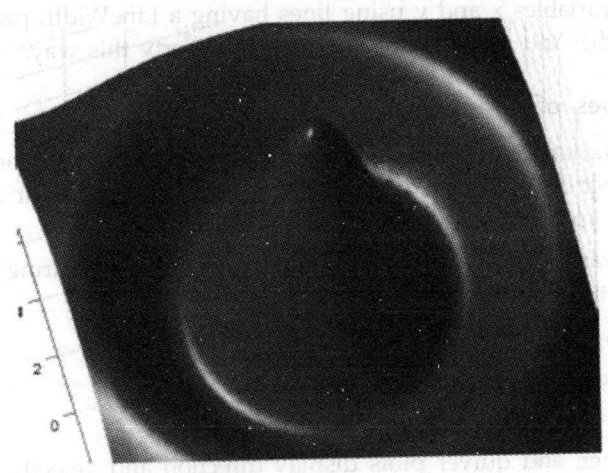

17. IMAGES

Two-dimensional arrays can be displayed as *images*, where the array elements determine brightness or color of the images. For example, the statements

```
>> load durer
>> whos
   Name      Size       Bytes      Class
   X         648x509    2638656    double array
   caption   2x28       112        char array
   map       128x3      3072       double array
```

load the file durer.mat, adding three variables to the workspace. The matrix X is a 648-by-509 matrix and map is a 128-by-3 matrix that is the colormap for this image.

18. HANDLE GRAPHICS

MATLAB creates the graph using various graphics objects, such as lines, text, and. All graphics objects have properties that control the appearance and behavior of the object. MATLAB enables to query the value of each property and set the value of most properties. Whenever MATLAB creates a graphics object, it assigns an identifier (called a handle) to the object. You can use this handle to access the object's properties. Handle Graphics is useful if you want to:

- Modify the appearance of graphs.
- Create custom plotting commands by writing M-files that create and manipulate objects directly.

18.1 Setting Properties from Plotting Commands

Plotting commands that create lines or surfaces enable you to specify property name/property value pairs as arguments. For example, the command

```
plot(x,y,'LineWidth',1.5)
```

plots the data in the variables x and y using lines having a LineWidth property set to 1.5 points (one point = 1/72 inch). You can set any line object property this way.

18.2 Different Types of Graphs

MATLAB supports a variety of graph types that enable you to present information effectively. The type of graph you select depends, to a large extent, on the nature of your data. The following list can help to select the appropriate graph:

- Bar and area graphs are useful to view results over time, comparing results, and displaying individual contribution to a total amount.
- Pie charts show individual contribution to a total amount.
- Histograms show the distribution of data values.
- Stem and stairstep plots display discrete data.
- Compass, feather, and quiver plots display direction and velocity vectors.
- Contour plots show equivalued regions in data.
- Interactive plotting enable you to select data points to plot with the pointer.
- Animations add an addition data dimension by sequencing plots.

18.2.1 Bar and Area Graphs

Bar and area graphs display vector or matrix data. These types of graphs are useful for viewing results over a period of time, comparing results from different datasets, and showing how individual elements contribute to an aggregate amount. Bar graphs are suitable for displaying discrete data, whereas area graphs are more suitable for displaying continuous data.

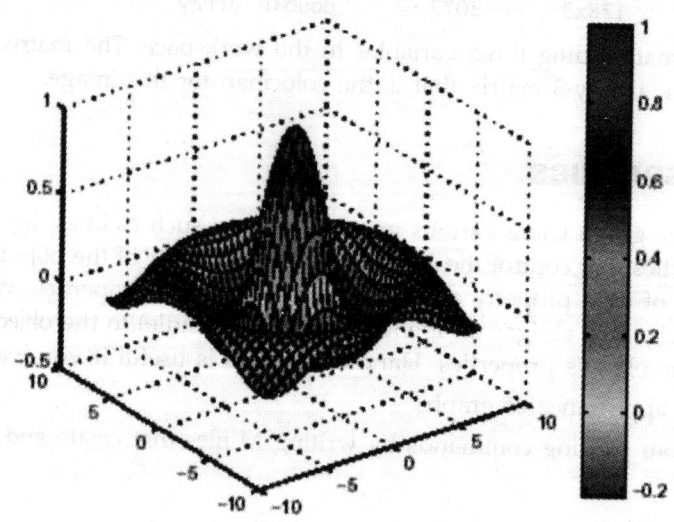

Function	Description
bar	Displays columns of m-by-n matrix as m groups of n vertical bars
barh	Displays columns of m-by-n matrix as m groups of n horizontal bars
bar3	Displays columns of m-by-n matrix as m groups of n vertical 3-D bars

| bar3h | Displays columns of m-by-n matrix as m groups of n horizontal 3-D bars |
| area | Displays vector data as stacked area plots graphs. |

Types of Bar Graphs: MATLAB has four specialized functions that display bar graphs. These functions display 2- and 3-D bar graphs, and vertical and horizontal bar

| Two-Dimensional | Three-Dimensional | Vertical | Horizontal |
| bar | bar3 | barh | bar3h |

Grouped Bar Graph: By default, a bar graph represents each element in a matrix as one bar. Bars in a 2-D bar graph, created by the bar function, are distributed along the x-axis with each element in a column drawn at a different location. All elements in a row are clustered around the same location on the x-axis.

Pie Charts: Pie charts display the percentage that each element in a vector or matrix contributes to the sum of all elements. pie and pie3 create 2-D and 3-D pie charts.

Example–Pie Chart: Here is an example using the pie function to visualize the contribution that three products make to total sales. Given a matrix X where each column of X contains yearly sales figures for a specific product over a five-year period,

```
X = [19.3 22.1 51.6; 34.2 70.3 82.4; 61.4 82.9 90.8; 50.5 54.9 59.1; 29.4 36.3 47.0];
```

sum each row in X to calculate total sales for each product over the five-year period.

```
x = sum(X);
```

You can offset the slice of the pie that makes the greatest contribution using the explode input argument. This argument is a vector of zero and nonzero values. Nonzero values offset the respective slice from the chart.

First, create a vector containing zeros of same size as X.

```
explode = zeros(size(x));
```

Then find the slice that contributes the most and set the corresponding explode element to 1.

```
[c,offset] = max(x);
explode(offset) = 1;
```

The explode vector contains the elements [0 0 1]. To create the exploded pie chart, use the statement.

```
h = pie(x,explode); colormap summer
```

Removing a Piece from a Pie Charts: When the sum of the elements in the first input argument is equal to or greater than 1, pie and pie3 normalize the values. So, given a vector of elements x, each slice has an area of $x_i/sum(x_i)$, where x_i is an element of x. The normalized value specifies the fractional part of each pie slice.

When the sum of the elements in the first input argument is less than 1, pie and pie3 do not normalize the elements of vector x. They draw a partial pie.

For example,

```
x = [.19 .22 .41];
pie(x)
```

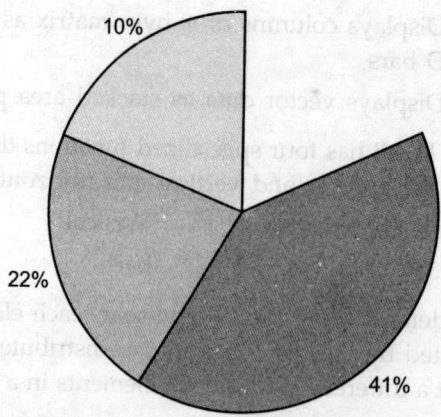

Histograms: MATLAB's histogram functions show the distribution of data values. The functions that create histograms are hist and rose.

Function	Description
hist	Displays data in a Cartesian coordinate system
rose	Displays data in a polar coordinate system

The histogram functions count the number of elements within a range and display each range as a rectangular bin. The height (or length when using rose) of the bins represents the number of values that fall within each range.

Histograms in Cartesian Coordinate Systems: The hist function shows the distribution of the elements in Y as a histogram with equally spaced bins between the minimum and maximum values in Y. If Y is a vector and is the only argument, hist creates up to 10 bins. For example,

```
yn = randn(10000,1);
hist(yn)
```

generates 10,000 random numbers and creates a histogram with 10 bins distributed along the x-axis between the minimum and maximum values of yn.

A Typical 3-D Graph: This table illustrates typical steps involved in producing 3-D scenes containing either data graphs or models of 3-D objects. Example applications include pseudocolor surfaces illustrating the values of functions over specific regions and objects drawn with polygons and colored with light sources to produce realism.

Step	Typical Code
1. Prepare your data	Z = peaks(20);
2. Select window and position subplot(2,1,2)	figure(1) plot region within window
3. Call 3-D graphing function	h = surf(Z);
4. Set colormap and shading set(h,'EdgeColor','k')	colormap hot algorithm shading interp
5. Add lighting light('Position',[-2,2,20]) lighting phong material([0.4,0.6,0.5,30]) set(h,'FaceColor',[0.7 0.7 0],... 'BackFaceLighting','lit')	

6. Set viewpoint
    view([30,25]) set(gca,'CameraViewAngleMode','Manual')

7. Set axis limits and tick marks
    axis([5 15 5 15 8 8]) set(gca'ZTickLabel','Negative | | Positive')

8. Set aspect ratio   set(gca,'PlotBoxAspectRatio',[2.5 2.5 1])

9. Annotate the graph with axis
    xlabel('X Axis') labels, legend, and text
    ylabel('Y Axis')
    zlabel('Function Value')
    title('Peaks')

10. Print graph   set(gcf,'PaperPositionMode','auto')
    print dps2

Line Plots of 3-D Data: The 3-D analog of the plot function is plot3. If x, y, and z are three vectors of the same length, plot3(x,y,z) generates a line in 3-D through the points whose coordinates are the elements of x, y, and z and then produces a 2-Dprojection of that line on the screen. For example, these statements produce a helix.

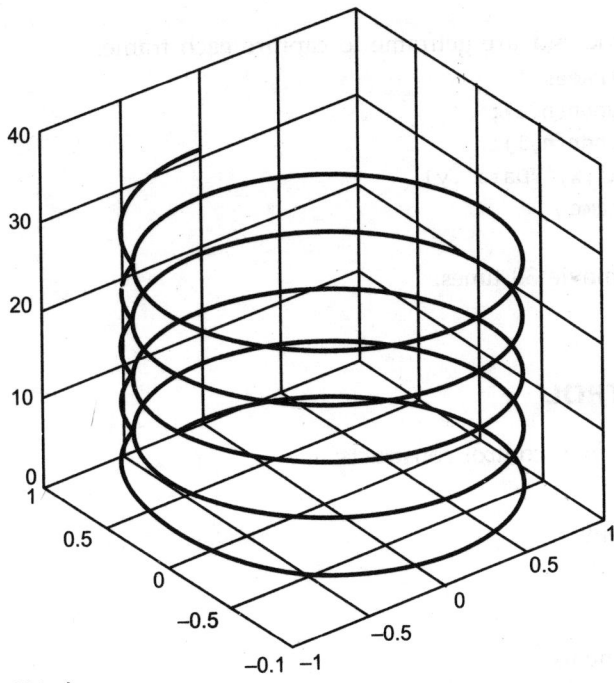

```
t = 0:pi/50:10*pi;
plot3(sin(t),cos(t),t)
axis square; grid on
```

19. ANIMATIONS

MATLAB provides two ways of generating moving, animated graphics :

- Continually erase and then redraw the objects on the screen, making incremental changes with each redraw.
- Save a number of different pictures and then play them back as a movie.

20. CREATING MOVIES

If you increase the number of points in the Brownian motion example to something like n = 300 and s = .02, the motion is no longer very fluid; it takes too much time to draw each time step. It becomes more effective to save a predetermined number of frames as bitmaps and to play them back as a *movie*. First, decide on the number of frames, say

```
nframes = 50;
```

Next, set up the first plot as before, except using the default EraseMode (normal).

```
x = rand(n,1)-0.5;
y = rand(n,1)-0.5;
h = plot(x,y,'.');
set(h,'MarkerSize',18);
axis([-1 1 -1 1])
axis square
grid off
```

Generate the movie and use getframe to capture each frame.

```
for k = 1:nframes
x = x + s*randn(n,1);
y = y + s*randn(n,1);
set(h,'XData',x,'YData',y)
M(k) = getframe;
end
```

Finally, play the movie 30 times.

```
movie(M,30)
```

21. FLOW CONTROL

MATLAB has several flow control constructs:

- if statements
- switch statements
- for loops
- while loops
- continue statements
- break statements

21.1 if statements

The if statement evaluates a logical expression and executes a group of statements when the expression is *true*. The optional elseif and else keywords provide for the execution of alternate groups of statements. An end keyword, which matches the if, terminates the last group of statements. The groups of statements are delineated by the four keywords – no braces or brackets are involved. MATLAB's algorithm for generating a magic square of order n involves three different cases: when n is odd, when n is even but not divisible by 4, or when n is divisible by 4. This is described by

```
if rem(n,2) ~= 0
M = odd_magic(n)
elseif rem(n,4) ~= 0
M = single_even_magic(n)
else
M = double_even_magic(n)
end
```

In this example, the three cases are mutually exclusive, but if they weren't, the first *true* condition would be executed. It is important to understand how relational operators and if statements work with matrices. When you want to check for equality between two variables, you might use

```
if A == B, ...
```

This is legal MATLAB code, and does what you expect when A and B are scalars. But when A and B are matrices, A == B does not test *if* they are equal, it tests *where* they are equal; the result is another matrix of 0's and 1's showing element-by-element equality. In fact, if A and B are not the same size,

```
then A == B is an error.
```

The proper way to check for equality between two variables is to use the *isequal* function,

```
if isequal(A,B), ...
```

21.2 switch and case Satements

The *switch* statement executes groups of statements based on the value of a variable or expression. The keywords case and otherwise delineate the groups. Only the first matching case is executed. There must always be an end to match the switch.

The logic of the magic squares algorithm can also be described by

```
switch (rem(n,4)==0) + (rem(n,2)==0)
   case 0
      M = odd_magic(n)
   case 1
      M = single_even_magic(n)
   case 2
      M = double_even_magic(n)
   otherwise
      error('This is impossible')
end
```

21.3 for Loops

The *for* loop repeats a group of statements a fixed, predetermined number of times. A matching end delineates the statements.

```
for n = 3:32
   r(n) = rank(magic(n));
end
r
```

The semicolon terminating the inner statement suppresses repeated printing, and there after the loop displays the final result. It is a good idea to indent the loops for readability, especially when they are nested.

```
for i = 1:m
    for j = 1:n
        H(i,j) = 1/(i+j);
    end
end
```

21.4 while Loops

The *while* loop repeats a group of statements an indefinite number of times under control of a logical condition. A matching end delineates the statements. Here is a complete program, illustrating while, if, else, and end, that uses interval bisection to find a zero of a polynomial. The program to simulated a given polynomial $x3 - 2x - 5$, as follows:

```
a = 0; fa = -Inf;
b = 3; fb = Inf;
while b-a > eps*b
    x = (a+b)/2;
    fx = x^3-2*x-5;
    if sign(fx) == sign(fa)
        a = x; fa = fx;
    else
        b = x; fb = fx;
    end
end
```

21.5 continue Statements

The *continue* statement passes control to the next iteration of the for or while loop in which it appears, skipping any remaining statements in the body of the loop. In nested loops, continue passes control to the next iteration of the for or while loop enclosing it. The example below shows a continue loop that counts the lines of code in the file, magic.m, skipping all blank lines and comments. A continue statement is used to advance to the next line in *magic.m* without incrementing the count whenever a blank line or comment line is encountered.

```
fid = fopen('magic.m','r');
count = 0;
while ~feof(fid)
    line = fgetl(fid);
    if isempty(line) | strncmp(line,'%',1)
        continue
    end
    count = count + 1;
end
disp(sprintf('%d lines',count));
```

21.6 break Statements

The break statement lets you exit early from a for or while loop. In nested loops, break exits from the innermost loop only. Here is an improvement on the example from the previous section. Why is this use of break a good idea?

```
a = 0; fa = -Inf;
b = 3; fb = Inf;
while b-a > eps*b
   x = (a+b)/2;
   fx = x^3-2*x-5;
   if fx == 0
      break
   elseif sign(fx) == sign(fa)
      a = x; fa = fx;
   else
      b = x; fb = fx;
   end
end
x
```

22. OTHER DATA STRUCTURES

This section introduces you to some other data structures in MATLAB, including:

- Multidimensional arrays
- Cell arrays
- Characters and text
- Structures

22.1 Multidimensional Arrays

Multidimensional arrays in MATLAB are arrays with more than two subscripts. They can be created by calling zeros, ones, rand, or randn with more than two arguments. For example,

```
R = randn(3,4,5);
```

creates a 3-by-4-by-5 array with a total of 3x4x5 = 60 normally distributed random elements. A three-dimensional array might represent three-dimensional physical data, say the temperature in a room, sampled on a rectangular grid. Or, it might represent a sequence of matrices, $A(k)$, or samples of a time-dependent matrix, $A(t)$. In these latter cases, the (i, j)th element of the kth matrix, or the tkth matrix, is denoted by $A(i, j, k)$. MATLAB's and Dürer's versions of the magic square of order 4 differ by an interchange of two columns.Many different magic squares can be generated by interchanging columns. The statement

```
p = perms(1:4);
```

generates the 4! = 24 permutations of 1:4. The kth permutation is the row

```
vector, p(k,:). Then
A = magic(4);
M = zeros(4,4,24);
for k = 1:24
M(:,:,k) = A(:,p(k,:));
end
```

stores the sequence of 24 magic squares in a three-dimensional array, M. The size of M is

```
size(M)
ans =
    4 4 24
```

It turns out that the third matrix in the sequence is Dürer's.

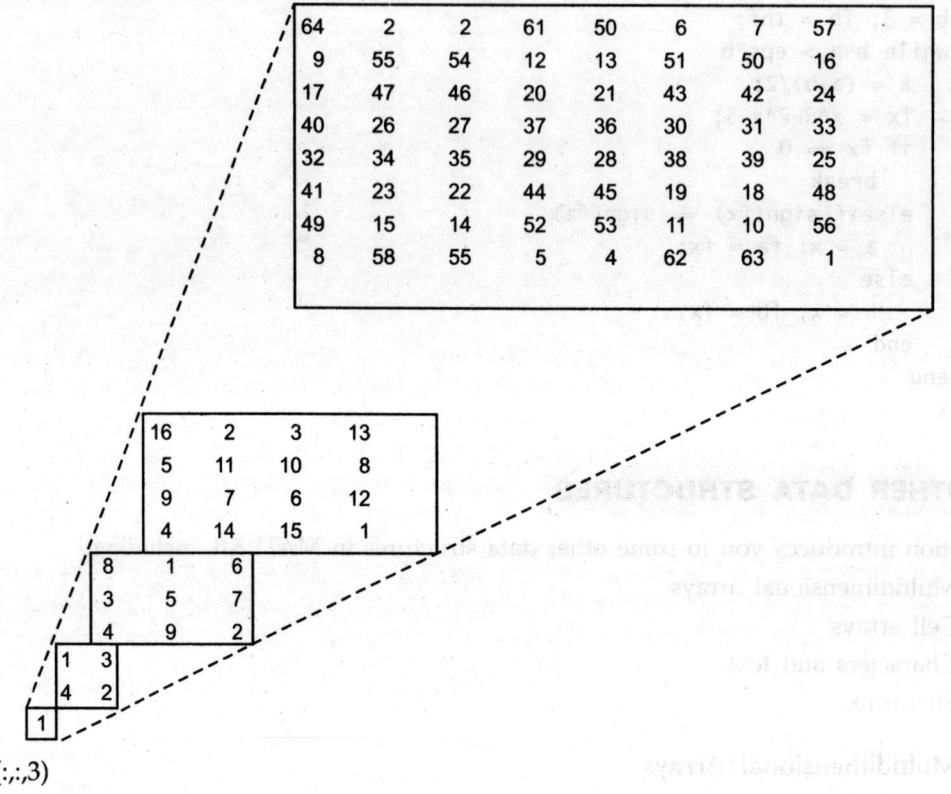

64	2	2	61	50	6	7	57
9	55	54	12	13	51	50	16
17	47	46	20	21	43	42	24
40	26	27	37	36	30	31	33
32	34	35	29	28	38	39	25
41	23	22	44	45	19	18	48
49	15	14	52	53	11	10	56
8	58	55	5	4	62	63	1

16	2	3	13
5	11	10	8
9	7	6	12
4	14	15	1

8	1	6
3	5	7
4	9	2

1	3
4	2

1

M(:,:,3)

```
         16  3   2  13
          5 10  11   8
ans =
          9  6   7  12
          4 15  14   1
```

The statement

```
sum(M,d)
```

computes sums by varying the dth subscript. So

```
sum(M,1)
```

is a 1-by-4-by-24 array containing 24 copies of the row vector

34 34   34   34

Finally,

```
S = sum(M,3)
```

adds the 24matrices in the sequence. The result has size 4-by-4-by-1, so it looks like a 4-by-4 array.

$$S = \begin{matrix} 204 & 204 & 204 & 204 \\ 204 & 204 & 204 & 204 \\ 204 & 204 & 204 & 204 \\ 204 & 204 & 204 & 204 \end{matrix}$$

22.2 Cell Arrays

Cell arrays in MATLAB are multidimensional arrays whose elements are copies of other arrays. A cell array of empty matrices can be created with the cell function. But, more often, cell arrays are created by enclosing a miscellaneous collection of things in curly braces, {}. The curly braces are also used with subscripts to access the contents of various cells. For example,

```
C = {A sum(A) prod(prod(A))}
```

produces a 1-by-3 cell array. The three cells contain the magic square, the row vector of column sums, and the product of all its elements.When C is displayed, you see

```
C = [4 x 4 double] [1 x 4 double] [20922789888000]
```

This is because the first two cells are too large to print in this limited space, but the third cell contains only a single number, 16!, so there is room to print it. Here are two important points to remember. First, to retrieve the contents of one of the cells, use subscripts in curly braces. For example, C{1} retrieves the magic square and C{3} is 16!. Second, cell arrays contain *copies* of other arrays, not *pointers* to those arrays. If you subsequently change A, nothing happens to C. Three-dimensional arrays can be used to store a sequence of matrices of the *same* size. Cell arrays can be used to store a sequence of matrices of *different* sizes. For example,

```
M = cell(8,1);
for n = 1:8
M{n} = magic(n);
end
M
```

produces a sequence of magic squares of different order.

```
M =    [ 1]
       [ 2x2 double]
       [ 3x3 double]
       [ 4x4 double]
       [ 5x5 double]
       [ 6x6 double]
       [ 7x7 double]
       [ 8x8 double]
```

You can retrieve our old friend with

```
M{4}
```

22.3 Characters and Text

Enter text into MATLAB using single quotes. For example,

```
s = 'Hello'
```

The result is not the same kind of numeric matrix or array we have been dealing with up to now. It is a 1-by-5 character array.

Internally, the characters are stored as numbers, but not in floating-point format. The statement

```
a = double(s)
```

converts the character array to a numeric matrix containing floating-point representations of the ASCII codes for each character. The result is

    a = 72 101 108 108 111

The statement

    s = char(a)

reverses the conversion. Converting numbers to characters makes it possible to investigate the various fonts available on your computer. The printable characters in the basic ASCII character set are represented by the integers 32:127. (The integers less than 32 represent nonprintable control characters.) These integers are arranged in an appropriate 6-by-16 array with

    F = reshape(32:127,16,6)';

The printable characters in the extended ASCII character set are represented by F+128. When these integers are interpreted as characters, the result depends on the font currently being used. Type the statements

    char(F)
    char(F+128)

and then vary the font being used for the MATLAB Command Window. Select **Preferences** from the **File** menu. Be sure to try the **Symbol** and **Wingdings** fonts, if you have them on your computer. Here is one example of the kind of output you might obtain.

Concatenation with square brackets joins text variables together into larger strings. The statement

    h = [s, ' world']

joins the strings horizontally and produces

    h = Hello world

The statement

    v = [s; 'world']

joins the strings vertically and produces

    v = Hello world

Note that a blank has to be inserted before the 'w' in h and that both words in v have to have the same length. The resulting arrays are both character arrays;

    h is 1-by-11 and v is 2-by-5.

To manipulate a body of text containing lines of different lengths, you have two choices – a padded character array or a cell array of strings. The char function accepts any number of lines, adds blanks to each line to make them all the same length, and forms a character array with each line in a separate row. For example,

    S = char('A','rolling','stone','gathers','momentum.')

produces a 5-by-9 character array.

    S =
    A
    rolling
    stone
    gathers
    momentum.

There are enough blanks in each of the first four rows of S to make all the rows the same length. Alternatively, you can store the text in a cell array. For example,

    C = {'A';'rolling';'stone';'gathers';'momentum.'}

is a 5-by-1 cell array.

```
C =
    'A'
    'rolling'
    'stone'
    'gathers'
    'momentum.'
```

You can convert a padded character array to a cell array of strings with

```
C = cellstr(S)
```

and reverse the process with

```
S = char(C)
```

Structures: Structures are multidimensional MATLAB arrays with elements accessed by textual *field designators*. For example,

```
S.name = 'Ram';
S.score = 83;
S.grade = 'B+'
```

creates a scalar structure with three fields.

```
S =
    name: 'Ram'
    score: 83
    grade: 'B+'
```

Like everything else in MATLAB, structures are arrays, so you can insert additional elements. In this case, each element of the array is a structure with several fields. The fields can be added one at a time,

```
S(2).name = 'Shyam';
S(2).score = 91;
S(2).grade = 'A-';
```

or, an entire element can be added with a single statement.

```
S(3) = struct('name','Rohan',...
'score',70,'grade','C')
```

Now the structure is large enough that only a summary is printed.

```
S =
    1x3 struct array with fields:
    name
    score
    grade
```

There are several ways to reassemble the various fields into other MATLAB arrays. They are all based on the notation of a *comma separated list*.

```
S.score
```

it is the same as typing

```
S(1).score, S(2).score, S(3).score
```

This is a comma separated list. Without any other punctuation, it is not very useful. It assigns the three scores, one at a time, to the default variable ans and dutifully prints out the result of each assignment. But when you enclose the expression in square brackets,

```
[S.score]
```

it is the same as

```
[s(1).score, s(2).score, s(3).score]
```

which produces a numeric row vector containing all of the scores.

```
ans =
        83 91 70
```

Similarly, typing

```
S.name
```

just assigns the names, one at time, to ans. But enclosing the expression in curly braces,

```
{s.name}
```

creates a 1-by-3 cell array containing the three names.

```
ans =
        'Ram'  'Shyam'  'Rohan'
And
        char(S.name)
```

calls the char function with three arguments to create a character array from the name fields,

```
ans =
        Ram
        Shyam
        Rohan
```

23. SCRIPTS AND FUNCTIONS

MATLAB is a powerful programming language as well as an interactive computational environment. Files that contain code in the MATLAB language are called M-files. You create M-files using a text editor, then use them as you would any other MATLAB function or command. There are two kinds of M-files :

- Scripts, which do not accept input arguments or return output arguments. They operate on data in the workspace.

- Functions, which can accept input arguments and return output arguments. Internal variables are local to the function. For new MATLAB programmer, just create the M-files in the current directory. If more number of m-files then it is necessary to organize them into other directories and add that to MATLAB's search path. If the m-file names are available in the directory, MATLAB executes the one that occurs first in the search path.

23.1 Scripts

To invoke a *script*, MATLAB simply executes the commands found in the m-file. Scripts can operate on existing data in the workspace, or they can create new data on which to operate. Although scripts do not return output arguments, any variables that they create remain in the workspace, to be used in subsequent computations. In addition, scripts can produce graphical output using functions like plot. For example, create a file called rank_A.m that contains these MATLAB commands.

```
% find the rank of a matrix
    A=5*eye(4);
    Rank_A=rank(A);
```

The following statement on Matlab command window to execute these commands.

```
>> r
```

After execution of the file is complete, the variables A and r remain in the workspace.

23.2 Functions

Functions are M-files that can accept input arguments and return output arguments. The name of the M-file and of the function should be the same. Functions operate on variables within their own workspace which is separate from the workspace used by MATLAB command prompt. In the above example theM-file rank.m is available in the directory toolbox/matlab/matfun and contains following commands.

```
function r = rank(A,tol)
    s = svd(A);
    if nargin==1
        tol = max(size(A)') * max(s) * eps;
    end
    r = sum(s > tol);
```

The first line of a function starts with the keyword function. It gives the function name and order of arguments. In this case, there are up to two input arguments and one output argument. The variable s introduced in the body of the function, as well as the variables on the first line, r, A and tol, are all *local* to the function. This example illustrates one aspect of MATLAB functions that is not ordinarily found in other programming languages – a variable number of arguments. The rank function can be used in several different ways.

```
>> rank(A)
ans =
    r = rank(A)
    r = rank(A,1.e-6)
```

Global Variables: If you want more than one function to share a single copy of a variable, simply declare the variable as global in all the functions. Do the same thing at the command line if you want the base workspace to access the variable. The global declaration must occur before the variable is actually used in a function. Although it is not required, using capital letters for the names of global variables helps distinguish them from other variables. For example, create an M-file called falling.m.

```
function h = force_m(acc)
global mass
h = mass*acc;
```

when we write the following commands on the command prompt

```
>> global mass
>> mass = 2; % kg
>> f = force((0:.1:5)');
```

The two global statements make the value assigned to mass at the command prompt available inside the function.

The Eval Function: The eval function works with text variables to implement a powerful text macro facility. The expression or statement eval(s) uses the MATLAB interpreter to evaluate the expression or execute the statement contained in the text string s. The example of the previous section could also be done with the following code, although this would be somewhat less efficient because it involves the full interpreter, not just a function call.

```
for d = 1:31
   s = ['load August' int2str(d) '.dat'];
   eval(s)
   % Process the contents of the d-th file
end
```

Vectorization: To obtain the most speed out of MATLAB, it's important to vectorize the algorithms in m-files.Where other programming languages might use FOR or DO loops, MATLAB can use vector or matrix operations. A simple example involves creating a table of logarithms.

```
x = .01;
for k = 1:1001
   y(k) = log10(x);
   x = x + .01;
end
```

A vectorized version of the same code is

```
x = .01:.01:10;
y = log10(x);
```

For more complicated code, vectorization options are not always so obvious. When speed is important, however, you should always look for ways to vectorize your algorithms.

Preallocation: If you can't vectorize a piece of code, you can make your for loops go faster by preallocating any vectors or arrays in which output results are stored. For example, this code uses the function zeros to preallocate the vector created in the for loop. This makes the for loop execute significantly faster.

```
r = zeros(32,1);
for n = 1:32
   r(n) = rank(magic(n));
end
```

Without the preallocation in the previous example, the MATLAB interpreter enlarges the r vector by one element each time through the loop. Vector preallocation eliminates this step and results in faster execution.

Function Functions: A class of functions, called "function functions," works with nonlinear functions of a scalar variable. That is, one function works on another function. The function functions include :

- Zero finding
- Optimization
- Quadrature
- Ordinary differential equations

MATLAB represents the nonlinear function by a function M-file. For example, here is a simplified version of the function humps from the matlab/demos directory.

```
function y = humps(x)
y = 1./((x-.3).^2 + .01) + 1./((x-.9).^2 + .04) - 6;
```

Evaluate this function at a set of points in the interval $0 \le x \le 1$ with

```
x = 0:.002:1;
y = humps(x);
```

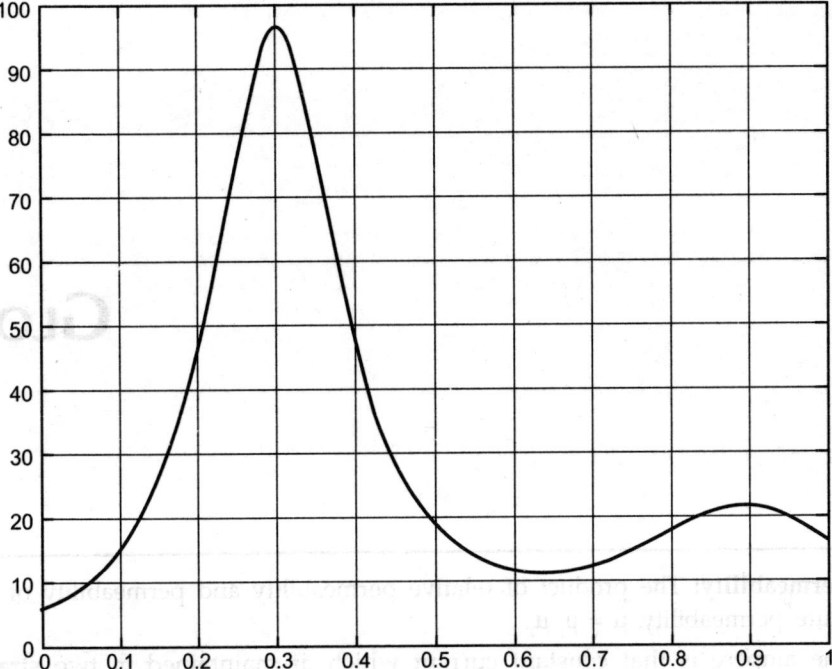

Then plot the function with

```
plot(x,y)
```

The graph shows that the function has a local minimum near $x = 0.6$. The function fminsearch finds the *minimizer*, the value of x where the function takes on this minimum. The first argument to fminsearch is a function handle to the function being minimized and the second argument is a rough guess at the location of the minimum.

```
p = fminsearch(@humps,.5)
p =
    0.6370
```

To evaluate the function at the minimizer,

```
humps(p)
ans =
    11.2528
```

Numerical analysts use the terms *quadrature* and *integration* to distinguish between numerical approximation of definite integrals and numerical integration of ordinary differential equations. MATLAB's quadrature routines are quad and quadl. The statement

```
Q = quadl(@humps,0,1)
```

computes the area under the curve in the graph and produces

```
Q =
    29.8583
```

Finally, the graph shows that the function is never zero on this interval. So, if you search for a zero with

```
z = fzero(@humps,.5)
```

you will find one outside of the interval

```
z =
```

GLOSSARY

A

Absolute permeability: The product of relative permeability and permeability of free space is called absolute permeability. $\mu = \mu_o \mu_r$

Ampere: The ampere is that constant current which, if maintained in two straight parallel conductors of infinite length, of negligible circular cross-section, and placed 1 meter apart in vacuum, would produce between these conductors a force equal to 2×10^{-7} newton per meter of length (9th CGPM, 1948).

AC circuit: electrical network in which the voltage polarity and directions of current flow change continuously, and often periodically.

AC coupling: a method of connecting two circuits that allows displacement current to flow while preventing conductive currents. Reactive impedance devices (*e.g.*, capacitors and inductive transformers) are used to provide continuity of alternating current flow between two circuits while simultaneously blocking the flow of direct current.

AC motor an electromechanical system that converts alternating current electrical power into mechanical power.

Air core transformer: two or more coils placed so that they are linked by the same flux with an air core. With an air core the flux is not confined.

Angular frequency: the rate of change of the phase of a wave in radians per second.

Armature reaction: (*a*) in DC machines, a distortion of the field flux caused by the flux created by the armature current. Armature reaction in a DC machine causes lower flux at one pole-tip and higher flux at the other, which may lead to magnetic saturation. It also shifts the neutral axis, causing sparking on the commutator.

(*b*) in AC synchronous machines, a voltage "drop" caused by the armature current. In the steady state model of the synchronous machine, the armature reaction is accounted for by a component of the synchronous reactance.

Average power: the average value, taken over an interval in time, of the instantaneous power. The time interval is usually one period of the signal.

B

Breakdown strength: Voltage gradient at which the molecules of medium break down to allow passage of damaging levels of electric current.

Balanced load: A load on a multi-phase power line in which each line conductor sees the same impedance.

Base speed: Corresponds to speed at rated torque, rated current, and rated voltage conditions at the temperature rise specified in the rating. It is the maximum speed at which a motor can operate under constant torque characteristics or the minimum speed to operate at rated power.

Blocked-rotor test: An induction motor test conducted with the shaft held so it cannot rotate. Typically about 25% of rated voltage is applied, often at reduced frequency and the current is measured. The results are used to determine the winding impedances referred to the stator.

Breakdown: As applied to insulation (including air), the failure of an insulator or insulating region to prevent conduction, typically because of high voltage.

Breakdown strength: Voltage gradient at which the molecules of medium break down to allow passage of damaging levels of electric current.

Breakdown torque: Maximum torque that can be developed by a motor operating at rated voltage and frequency without experiencing a significant and abrupt change in speed. Sometimes also called the stall torque or pull-out torque.

Brush: A conductor, usually carbon or a carbon–copper mixture, or graphite that makes sliding electrical contact to the rotor of an electrical machine. Brushes are used with slip rings on a synchronous machine to supply the DC field and are used with a commutator on a DC machine.

Bushing transformer: A potential transformer which is installed in a transformer bushing so as to take advantage of the insulating qualities of that bushing.

C

Controlling torque: The torque which controls the movement of the moving system and the pointer over the scale. It is usually provided by spring control or gravity control.

Capacitive reactance: The opposition offered to the flow of an alternating or pulsating current by capacitance measured in ohms.

Capacitance: The measure of the electrical size of a capacitor, in units of farads.

Coil: A conductor shaped to form a closed geometric path. Coils may have multiple turns.

Coil side: That portion of a motor or generator winding that cuts (or is cut by) lines of magnetic flux and, thus, contributes to the production of torque and Faraday EMF in the winding.

Coil span: The distance, measured either in number of coil slots or in spatial (mechanical) degrees, between opposite sides of a winding of an electric machine. A fullspan (full-pitch) winding is one in which the winding span equals the span between adjacent magnetic poles. Windings with span less than the distance between adjacent magnetic poles are called short-pitch, fractional-pitch, or chorded windings. Also called coil pitch.

Commutation: It is the process by which alternating current in the rotating coil of a DC machine is converted to unidirectional current.

Commutator: A cylindrical assembly of copper segments, insulated from each another by mica ᵗᵗ makes electrical contact with stationary brushes, to allow current to flow from the rotating armature windings of a DC machine to the external terminals of the machine. It also, enables reversal of current in the armature winding.

Conductance: The reciprocal of resistance.

Counter-EMF: A voltage developed in an electrical winding by Faraday's Law that opposes the source voltage, thus limiting the current in the winding.

Current transformer (CT): A transformer that is employed to provide a secondary current proportional to primary current flowing. It is used in current measurement, protective relays, and power metering applications.

Cylndrical-rotor alternator: A synchronous machine with a cylindrical rotor containing a distributed field winding. It is two or four pole machines (3000 and 1500 rpm at 50 Hz) and is usually used in large generators.

D

Damping torque: When the moving system of an instrument is subjected to deflecting and controlling torques, due to its inertia, the pointer oscilates around its final position before it settles down there. To damp out these oscillations some damping torque is required, which is provided by air friction, fluid friction or eddy currents.

Deflecting torque: The torque which cuases the moving system and the pointer of the instrument to get the required deflection.

Delta connection: The sources or loads in a three-phase system connected end-to-end, forming a closed path, like the Greek letter Δ.

Dielectric: Solid, liquid, or gaseous substance that acts as an insulation to the flow of electric current. A medium that exhibits negligible or no electrical conductivity and thus acts as a good electrical insulator. It is usually used to separate two conducting bodies to form a capacitor.

Damper winding: An uninsulated winding, embedded in the pole shoes of a synchronous machine, that includes several copper bars short-circuited by conducting rings at the ends, used to reduce fluctuation in the machine (speed fluctuation in motors and voltage fluctuations in generators).

Delta–delta transformer: A three-phase transformer connection formed by connecting three single-phase units in which the windings on both the primary and the secondary sides are connected in delta to form a closed path.

Delta-star transformer: Delta-star transformer is a three-phase transformer connection formed by connecting three single-phase transformers in which the primary windings are connected in delta to form a closed path while the secondary windings are connected in star and form a common point (the neutral).

Demagnetizing field: The magnetic field produced by the armature of dc machine which opposes the main magnetic field.

Diamagnetic materials with magnetization directed opposite to the magnetizing field, so that the permeability is less than one; metallic bismuth is an example.

Dielectric constant: a quantity that describes how a material stores and dissipates electrical energy.

E

Eddy current Loss: The emf induced in ferromagnetic material due to flux reversal, gives rise to eddy current. The product of the square of eddy current and resistance of specimen (core) gives rise to eddy current loss. $W_e = K_e f^2 B_m^2$ watts , where Bm—Maximum flux density and f—frequency of flux.

Electric braking: The reverse torque is produced based on electromagnetic principles is called electrical braking. In dc machines it is obtained by plugging, rheostatic braking, or regenetive braking.

Electric vehicle: Vehicle using electric energy storage, electric controls, and electric propulsion devices.

Eddy current: A circulating current in magnetic materials that is produced as a result of time-varying flux passing through a metallic magnetic material.

Eddy current brake: A braking device in which energy is dissipated as heat by generating eddy currents.

Eddy current drive: A magnetic drive coupled by eddy currents induced in an electrically conducting member by a rotating permanent magnet, resulting in a torque that is linearly proportional to the slip speed.

Effective length: The ratio of the voltage induced across an antenna terminating impedance divided by the incident field strength.

Electric current density: A source vector in electromagnetics that quantifies the amount of electric charge crossing some cross–sectional area per unit time. The direction of the electric current density is in the direction of electric charge motion. SI units are amperes per square meter.

Electric permittivity: Tensor relationship between the electric field vector and the electric displacement vector in a medium with no hysteresis; displacement divided by the electric field in scalar media.

Equivalent circuit: A combination of electric circuit elements chosen to represent the performance of a machine or device by establishing the same relationships for voltage, current, and power.

Exciter: A DC source that supplies the field current to produce a magnetic flux in an electric machine. Often it may be a small DC generator, placed on the same shaft of the electrical machine.

F

Form Factor: It is the ratio of rms value to average value. $K_f = \dfrac{\text{rms value}}{\text{average value}}$

Ferrite: A term applied to a large group of ceramic ferromagnetic materials usually consisting of oxides of magnesium, iron, and manganese. Ferrites are characterized by permeability values in the thousands and are used for RF transformers and high Q coils.

Ferrite core: A magnetic core made up of ferrite (compressed powdered ferrromagnetic) material, having high resistivity and low eddy current loss.

Ferromagnetic materials: in which internal magnetic moments spontaneously line up parallel to each other to form domains, resulting in permeabilities considerably higher than unity (in practice, 1.1 or more); for examples iron, nickel, and cobalt.

Flat-compounded DC generator: It designs in which the output voltage is maintained essentially constant over the entire range of load currents.

Flux: Lines that indicate the intensity and direction of a field. Intensity is usually represented by the density of the lines.

Flux density: Lines of magnetic flux per unit area, measured in tesla; $1 \text{Tesla} = 1 Wb/m^2$.

Flux linkage: Quantity that indicates the amount of flux associated with a coil.

Four-point starter: A manual motor starter that requires a fourth terminal for the holding coil. Because of its independent holding coil circuit, it is possible to vary the current in the field circuit independently of the holding coil circuit. The disadvantage is that the motor starter holding relay will not drop out with loss of the field; however, proper overcurrent protection should shut down the motor in the event of field loss.

Frequency: The repetition rate of a periodic signal.

G

General-purpose motor: These are the motors typically used when relatively low starting currents, low slip, good speed regulation, moderate starting torque, and high efficiency are the predominant concerns.

Generator: In electrical systems, any of a variety of electromechanical devices that convert mechanical power into electrical power, typically via Faraday induction effects between moving and stationary currentcarrying coils and/or magnets. Electrostatic generators use mechanical motion to physically separate stationary charges to produce a large electrostatic potential between two electrodes.

Ground: An earth-connected electrical conducting wire that may be designed to keep at zero potential all the time.

Ground current: The current that flows in a power system in a loop involving earth and (in some usages) other paths apart from the three phases.

H

Harmonic frequency: Integral multiples of fundamental frequency. For example, for a 50-Hz supply the harmonic frequencies are 150, 250, 350, ...

Hysteresis loss: The energy consumed by a specimen of a ferromagnetic material through a cycle of magnetization. $W_h = K_h f B_m^n$ watts

Harmonic frequency: Integral multiples of fundamental frequency. For example, for a 50-Hz supply, the harmonic frequencies are 150, 250, ...

Hydroelectric generator: Large, threephase synchronous alternator powered by a water-driven turbine.

Hysteresis curve: A graph describing the relationship between the magnetic flux density and the magnetic field intensity in a (usually ferromagnetic) material.

I

Inductor: A conductor used to introduce inductance into a circuit.

Ideal transformer: A transformer with zero winding resistance and a lossless, infinite permeability core resulting in a transformer efficiency of 100 percent. Infinite permeability would result in zero exciting current and no leakage flux. For an ideal transformer, the ratio of the voltages on the primary and secondary sides would be exactly the same as the ratio of turns in the windings, while the ratio of currents.

Inductance: A parameter that describes the ability of a device to store magnetic flux.

Induced voltage: Voltage produced by a time-varying magnetic flux linkage.

Insulator: A device designed to separate and prevent the flow of current between conductors. Properties of the dielectric (insulating) material and geometry of the insulator determine maximum voltage and temperature ratings.

Interpole: A set of small poles located midway between the main poles of a DC machine, containing a winding connected in series with the armature circuit. The interpole improves commutation by neutralizing the flux distortion in the neutral plane caused by armature reaction.

Isolation transformer: A transformer, typically with a turns ratio of 1:1, designed to provide galvanic isolation between the input and the output.

L

Leakage current: Stray direct current of relatively small value which flows through a capacitor when voltage is impressed across it.

Lamination: A thin sheet of metal used to build up the core of an electromagnetic device. Laminations are insulated from each other to reduce the losses associated with eddy currents.

Lap winding: An armature winding on a DC machine in which the two ends of each coil are connected to adjacent bars on the commutator ring. The lap winding provides "P"parallel paths through the armature winding, where P is the number of poles in the machine.

Lightning arrestor: A voltage-dependent resistor which is connected in parallel with lightning-susceptible electrical equipment. It provides a low-resistance electrical path to ground during overvoltage conditions, thus diverting destructive lightning energy around the protected equipment.

Load balancing: The process of trying to distribute work evenly among multiple computational resources.

Load tap changer (LTC): A tapped transformer winding combined with mechanically or electronically switched taps that can be changed under load conditions. The load tap changer is used to automatically regulate the output of a transformer secondary as load and source conditions vary.

Lad torque: The resisting torque applied at the motor shaft by the mechanical load that counterbalances the shaft torque generated by the motor and available at the shaft.

M

Magnetic field: Magnetic force field where lines of magnetism exist.

Magnetic Flux Density: Magnetic flux density or magnetic inductance is defined as the magnetic flux per unit area at right angles to the direction of the flux. Its unit is tesla or weber/m^2.

Magnetic flux: Term for lines of forces of magnetism. The lines of magnetic flux never intersect each other and they behave like stretched elastic cords, always try to shorten themselves.

Magnetic losses: Whenever magnetic materials subjected to periodic flux reversal, invariably hysterisis and eddy current losses occur called magnetic or iron losses.

Magnetomotive force: The force by which the magnetic field is produced, either by a current flowing through a coil of wire or by the proximity of a magnetized body. The amount of magnetism produced in the first method is proportional to the current through the coil and the number of turns in it.

Measuring Instruments: The devices used for measurements of electrical quantities like voltage, current, power, energy, resistance, etc.

Mutual inductance: The property that exists between two current-carrying conductors when the magnetic lines of force from one link with those from another.

MTTF: The mean time to failure is the mean or expected value of "time to failure".

N

NaN: Acronym for not a number. Used in IEEE floating-point representations to designate values that are not infinity or zero or within the bounds of the representation.

Negative sequence: The set of balanced but reverse sequence (*acb*) components used in symmetrical component analysis.

Negative-sequence reactance: It is an inductive reactance offered by a circuit for the flow of negative-sequence currents alone.

No-load test: Measure voltage, current and power under no load situation of a machine. This test is used to determine the magnetizing reactance of the machine equivalent circuit.

No load tap changer: Device that provides for changing the tap position on a tapped transformer when the transformer is de-energized. Different taps provide a different turns ratio for the transformer.

O

Ohmic loss: The power loss in conductive medium.

Oil-filled transformer: A transformer in which the magnetic core and the windings are submerged in insulating oil. In this oil serves as an insulator as well as the cooling medium for the transformer.

Oil-paper insulation: An insulation scheme used in transformers and cables in which conductors are insulated with heavy paper impregnated with dielectric oil.

Open-delta transformer: A connection similar to a delta–delta connection, except that one single-phase transformer is removed. It is used to deliver three-phase power using only two single-phase transformers. The normal capacity of the open-delta transformer is reduced to 57.7% of its delta rating.

Overexcited: The condition where the field winding current is greater than a rated value.

Overload: A situation that results in electrical equipment carrying more than its rated current. In a generator excess electrical load and in motor too much mechanical load would cause an overload.

P

Peak Factor: It is the ratio of peak value to rms value. It is also called crest factor or the amplitude factor. For sinusoidal wave $Kp = \sqrt{2} = 1.414$, $K_p = \dfrac{\text{Peak/Maximum Value}}{\text{rms Value}}$.

Permeance: The reciprocal of reluctance is called permeance.

Permeability: This is the ability of the medium to set up a magnetic flux density by magnetizing force H. Absolute permeability or Magnetic permeability of air $\mu_o = 4\pi \times 10^{-7}$ H/m.

Phase: The angular relationship between current and voltage in an ac circuit.

Phasor: A complex number representing a sinusoid; its magnitude and angle are the rms value and phase of the sinusoid, respectively.

Power factor (PF): The ratio of effective resistance to impedance of a circuit.

Paramagnetic materials: The materials with permeability slightly greater than unity. Sodium, potassium, and oxygen are examples of paramagnetic material.

Q

Quality factor (Q): The ratio of the reactance to its equivalent series resistance.

Quadrature axis (q axis): An axis placed 90 degrees ahead of the direct axis of a synchronous machine.

Qadrature axis synchronous reactance: The sum of the stator winding leakage reactance and the q-axis magnetizing (armature) reactance of a synchronous machine.

Quadrature axis subtransient reactance: The reactance characterizes the equivalent reactance of the q-axis windings of the machine during the initial time following a system disturbance. It is the sum of the stator winding leakage reactance, and the parallel combination of the q-axis magnetizing reactance and the q-axis rotor amortisseur leakage reactances.

Quadrature axis transient reactance: A value that characterizes the equivalent reactance of the q-axis windings of the synchronous machine between the initial time following a system disturbance (subtransient interval) and the steady state.

R

Reactance (X): Opposition to the flow of alternating current. Capacitive reactance (Xc) is the opposition offered by capacitors at a specified frequency and is measured in ohms.

Reactive power: Consider an ac source connected at a pair of terminals to an otherwise isolated network. The reactive power is a measure of the energy exchanged between the source and the network without being dissipated in the network. The reactive power delivered would be expressed as $Q = V \cdot I \sin(\theta)$.

Real power: Consider an ac source connected at a pair of terminals to an otherwise isolated network. The real power, equal to the average power, is the power dissipated by the source in the network.

Regulation: The change in quantity from no-load to full-load expressed as a percentage of full-load quatity. Quantity may be voltage in case of transformers or generators or speed in motors.

Relative permeability: The ratio of flux density in medium or material to the flux density produced in vacuum by same magnetizing force is called relative permeability. $\mu_r = 1$ for non-magnetic materials, and $\mu_r = 1000-10000$ for magnetic materials

Reluctance: The opposition offered by a magnetic circuit to the flux is called reluctance. Its unit is AT/wb.

Resistivity: The resistance of a conductor with unit length and unit cross-sectional area.

Residual magnetism: A form of flux remaining in a ferromagnetic material after the MMF is removed.

Rotational loss: It is primarily due to the rotation of the rotating parts in a machine and include the friction and windage losses. Also called mechanical loss.

Rotor: The rotating part of an electrical machine including the shaft, such as the rotating armature of a DC machine or the field of a synchronous machine.

S

Star (Y) connection: The three sources or loads in a three-phase system connected to have one common point, like the letter Y.

System: A set of objects with relations between them and their attributes or properties. It is embedded in an environment containing other interrelated objects.

Saturation: In ferromagnetic circuits, the magnetic flux initially increases linearly with the applied magnetomotive force (MMF), but eventually most of the domains in the ferromagnetic material become aligned, and the rate of increase in flux decreases as the MMF continues to increase.

Skew: An arrangement of slots or conductors in squirrel cage rotors so that they are not parallel to the rotor axis.

Skin effect: The tendency of an alternating current to concentrate in the areas of lowest impedance.

Slot pitch: The angular distance (normallyin electrical degrees) between the axes of two slots.

Speed regulation the variation of the output speed of a machine as the load on the shaft is increased from zero to some specified fraction of the full load or rated load. Usually expressed as a percentage of the no-load speed. A large speed regulation is most often considered as a bad regulation from a control point of view.

Subtransient impedance the series impedance that a generator or motor exhibits during the subtransient period, typically the first few cycles of a fault.

T

Temperature coefficient of resistance: The change in electrical resistance of a resistor per unit change in temperature.

Time constant: In a capacitor-resistor circuit, the number of seconds required for the capacitor to reach 63.2% of its full charge after a voltage is applied. The time constant of a capacitor with a capacitance (C) in farads in series with a resistance (R) in ohms is equal to $R \times C$ seconds. For inductive circuits time constant is L/R. To overcome this loss the core is made by stacking of thin laminated steel sheets to increase the resistance.

Traction motor: Electric motor that provides motive power to move vehicles.

Torque angle: The displacement angle between the rotor and rotating magnetic flux of the stator due to increases in shaft load in a synchronous machine.

U

Unsymmetrical load: A load which forces the currents in the three-phase power line which supplies it to be unequal.

V

Voltage rating: The maximum voltage that may be applied to the component or equipment.

Variable loss: These are the losses which change with a change in load or load current. For example, in a transformer, the winding losses are a function of the load current, while the core losses are almost independent of the load current.

W

Winding: A conductive path, usually wire of aluminium or copper which is inductively coupled.

Winding factor: A design parameter for electric machines that is the product of the pitch factor and the distribution factor.

Chapman S.J., *Electric Machinery Fundamentals*, 3rd edition McGraw Hill Singapore, 1999.

Chaturvedi D.K., *Soft Computing Techniques and its Applications in Electrical Engineering*, Springer Verlag, Berlin, Germany, 2008.

Clayton A.E., *The Performance and Design of DC machines*, Pitman and Sons

Cotton H., *Advanced Electrical Technology*, Wheeler Publishing.

Del Toro V. *Principles of electrical Engineering*, Prentice hall International.

Dubey G.K., *Fundamentals of electrical Drives*, Narosa Publishing House, New Delhi, 1995.

Fitzgerlad AE, Kingsley C and Umans SD, *Electric Machinery*, Fifth Edition McGraw Hill, New York, 2002.

Hayt W.H. & Kennedy J.E., *Engineering circuit Analysis*, McGraw Hill.

Irving L.Kosow, *Electric Machines and Transformer*, Prentice Hall of India.

Kothari D.P. and Nagrath I.J., *Electrical Machines*, Tata McGraw Hill, New Delhi, 2004.

Langsdrof E.H., *Theory of alternating Current Machinery*, McGraw Hill, New York, 1955.

Matlab Manual ver. 7.0, Math Works.

Nasar A. Syed, *Electric Machines and Power System*, Vol. 1 Electric Machines, McGraw Hill, New York, 1995.

Ryff P.T., *Electric Machinery*, Prentice Hall, New Jersey, 1988.

Sag M.G., *Alternating Current Machines*, 5th Edition, Sir Isaac Pitman and Sons Ltd. 1983.

Smith I McKenzie, *Electrical Technology*, Addison Wesley, 1995.

Staff E.E., MIT, *Magnetic Circuits and Transformers*, John Wiley, New York, 1943.

Sankaran C., *Introduction to Transformers*, New York: IEEE Press, 1992.

Heydt G.T., *Electric Power Quality*, Stars in a Circle Publications, 1996.

Guru B.S. and Hirizoglu H.R., *Electric Machinery and Transformers*, Saunders, 1996.

Sawhney A.K., *Electrical Machines Design*, Dhanpat Rai & Co., New Delhi, 1998.